Wendy's Home for Lost Boys

A Rewoven Fairytale of Peter Pan

SARA MADDEN

United States of America
www.saramadden.com

ISBN 978-1-946505-18-7 (eBook)
ISBN 978-1-946505-17-0 (Paperback)
ISBN 978-1-946505-16-3 (Hard Copy)

Printed in the United States of America
10 9 8 7 6 5 4 3 2 1

Cover design by Sara Madden
Book design and interior by Sara Madden
Cover images @ Freepik
Author photograph @ Zaugg Studios

For the Lost Boys
and the mothers who kept the windows open
And for my Grandson, Peter—Nana loves you

"To die would be an awfully big adventure."
—J.M. Barrie, Peter Pan

"I think maybe the adventure isn't dying.
I think maybe the adventure is staying alive."
—Peter Panning

"Humankind has not woven the web of life. We are but one thread within it. Whatever we do to the web, we do to ourselves. All things are bound together. All things connect."
——Chief Seattle, Choctaw Tribe

Author's Note

This book began with a question I could not let go of: what happens to the children who never make it home—and to the women who wait for them?

Wendy's Home for Lost Boys is a reimagining of the *Peter Pan* myth, written not from the perspective of flight and fantasy but from the ground—where mothers watch windows, where grief does not resolve neatly, and where care becomes an act of resistance.

This story is rooted in history. Edwardian London was a city filled with lost children. Some were orphaned by poverty and illness, some were bruised by neglect or violence, and, later, some would be sent to war before they were old enough to understand what war would take from them. The figure of the "lost boy" was never a metaphor alone.

This novel does not seek to diminish the magic of childhood. Instead, it asks what that magic costs—and who is expected to pay for it. It is written with deep respect for J.M. Barrie's work, while also acknowledging the silences it leaves behind.

Some readers may find parts of this book difficult. Themes of loss, trauma, and abuse are present, though handled with care and restraint. I encourage readers to move through the story at their own pace and to step away when needed.

At its heart, this is a book about watching. About remembering. About the belief that love is not proven by letting go but by staying. By keeping the window open, even when the night is long.

This book is for the children who were lost. And for the ones who waited.

Prologue

The Window Between Worlds

France, November 11, 1918, 4:47 a.m.

The war would end in six hours and thirteen minutes.

Peter Panning did not know this yet. He knew only that he was cold, that his boots had not been dry in eleven days, and that somewhere in the darkness beyond the trench wall, German machine guns waited with the patience of wolves.

He knew that his hands were shaking. They had been shaking for months now. It was a fine tremor he could not control, could not hide, could not stop. The other boys pretended not to notice. That was the kindness of the trenches—you learned not to see the things that would break you if you looked too closely.

The ledger was open in his lap.

By now, he had filled three of leatherbound books with names, with stories, with the small details that made a boy a person instead of a casualty report. *Harrison, J. Seventeen. Liverpool. Wanted to be a teacher. Afraid of spiders. Died at Passchendaele, calling for his mother.*

Thousands of names. Thousands of boys who had flown toward thunder and never come back.

Peter traced the most recent entry with his finger. The ink was barely dry. *Tomkins, R. Sixteen. Sheffield. Lied about his age—said he was nineteen, looked about twelve. Good at cards. Terrible at lying, except to recruiters. Died November 10, 1918. Shell. Quick.*

Quick. As if that mattered. As if the speed of dying made the death smaller, easier, more bearable.

Nothing made it bearable.

In the darkness beside him, Charlie Morton stirred—*Slightly*, they called him now, because he was always slightly miserable, slightly sarcastic, slightly certain he was smarter than everyone else. He had been Peter's shadow for three years, through Ypres and the Somme and all the other names that had become synonyms for Hell.

"Can't sleep?" Slightly's voice was hoarse, barely a whisper.

"Can you?"

"No." A pause. "Heard a rumor."

"There's always a rumor."

"This one's different." Slightly shifted, and Peter heard the wet squelch of mud. "Runner came through an hour ago. Said the Armistice is signed. Said the war ends at eleven."

Peter went very still. The words didn't make sense. They were sounds without meaning, shapes without form. The war had been the only reality for so long that he had forgotten there could be anything else.

"Eleven," he repeated.

"Eleven o'clock. This morning." Slightly's voice cracked. "Six hours, Pan. Six hours and it's over."

Six hours.

Peter looked at the ledger in his hands. At the thousands of names. At all the boys who had almost made it, who had died days or weeks or hours before this moment, who would never know how close they had been.

Six hours.

He thought of his mother. He had been thinking of her more lately—not of specific memories, which had grown hazy with time and trauma—but the *feeling* of her. The warmth. The safety. The absolute certainty that someone in the world was waiting for him, watching for him, keeping a window open so he could find his way home.

She had told him once about light. How it passed through a prism and scattered into colors. How it didn't disappear, it just found new places to land.

He had been trying to be like that. To scatter the light of all these lost boys into the world so they wouldn't disappear, so they would land somewhere and matter.

Six hours.

"They'll send us over, won't they. One last push," Peter said quietly.

Slightly was silent. That was answer enough.

"The generals want a victory," Peter continued. "Something to put in the papers. 'British forces advance on final day of war.' They don't care how many of us—" He stopped. He couldn't finish.

Tick. Tick. Tick.

That was the noise that haunted him. Not the shells, not the screaming, not the wet terrible sounds of boys dying in the mud. The ticking. The crocodile from the story his mother used to tell—the beast that had swallowed a clock, that you could always hear coming, that never stopped hunting.

The war was the crocodile. And the clock inside it had been ticking for four years, counting down the boys, swallowing them one by one.

Six more hours of ticking.

Six more hours of teeth.

"I'm going to write while there's still time," Peter said. "Everyone in the unit. Everyone I can reach."

"Now?"

"When else?" He was already pulling out his pencil and turning to a fresh page. "If we go over at dawn... if some of us don't—"

Slightly's hand closed over his wrist. "Pan. We're going to make it."

"You don't know that."

"No. But I'm choosing to believe it." Slightly's grip tightened. "And you're going to believe it too. Because you're the one who taught me that believing matters. That names matter. That we're not just numbers, we're people, and people don't give up six hours before the end."

Peter looked at his friend—his brother, really, after everything they had survived together. Slightly's face was gaunt, his eyes hollow, his skin the gray-yellow of men who had lived too long underground. But beneath the exhaustion, something burned. Something that refused to go out.

Peter opened the ledger.

He began to write.

This is a story about lost boys.
About the ones who flew toward thunder and didn't come back. About the ones who came home carrying thunder inside them. About mothers who kept windows open through four years of war, watching for sons who might never return. About nurses and gravediggers and all the hidden caretakers who held the broken boys together when everything else was falling apart.

It is a story about names.
The ones written in ledgers, carved on crosses, and whispered in the dark by people who refused to let them be forgotten.

But before it is any of those things, it is a story about one boy. One window. One mother who believed that love was stronger than war and proved it by waiting.

Her name was Wendy.
His name was Peter.
And this is how they found each other again.

Part I

The Window

CHAPTER ONE

SNOWFALL AND COLLISION

London, December 1896

London liked to pretend at gentility in winter. Snow draped itself over the city like a lace shawl pulled tight around a shivering old woman, softening the soot-stained brick and the rough edges of chimney pots. Carriages moved slowly along the slush-slick streets, their wheels creaking, horses steaming like kettles. From a distance, it might be mistaken for a painting—quiet, refined, almost serene.

It was considerably less serene when one was late.

Wendy Smead pressed her gloved hand to the brim of her hat as she ran, boots sliding on the cobblestones, skirts hitched higher than her mother would have approved. Her satchel bounced against her side, heavy with an anatomy textbook and the small wrapped parcel she had meant to deliver fifteen minutes earlier.

"Excuse me—pardon me—so sorry!"

She darted around a cluster of shoppers huddled beneath an awning. Someone hissed something about young ladies and decorum, but Wendy only flushed and kept running.

She had promised Matron she would return to the hospital by three. She had promised her mother she would bring the Christmas violets to Aunt Edith. She had promised herself she would become the sort of nurse whose shoes were always polished, whose charting was always neat, whose hat was never, ever askew.

At the moment, her hat was decidedly askew. Snowflakes clung to her lashes as she squinted through the white blur. The omnibus stop waited just ahead—past the lamppost, past the crowd, past the—

She collided with something solid and warm and entirely unexpected.

The world became a whirl of gray wool and startled breath. A man's voice said "Oof!" as though she had knocked the wind from him, which she very likely had.

Wendy's hat flew from her head. Her satchel slipped from her shoulder. The wrapped parcel shot from its pocket and tumbled into the snow.

She flailed, boots skidding—

Two firm hands caught her elbows and held her steady.

"Careful," a voice said. Low, unhurried, with a faint lilt of Yorkshire beneath the London polish. "The pavement's more ice than road today."

Wendy blinked snowflakes from her eyes and looked up. The man was perhaps twenty-three or twenty-four, with dark blond hair already freckled with melting snow and spectacles slightly crooked on his nose. He had a thin, earnest face and the rumpled look of someone who owned more books than common sense. His coat was good but not new. His eyes, behind the spectacles, were the color of tea with too much milk. They were warm and slightly apologetic, as though he was already sorry for something he hadn't yet done.

"I am so terribly sorry," Wendy blurted.

She dropped into a half-curtsy on instinct, which nearly sent her sliding again. His hands tightened on her arms.

"Best not try that again," he said. She caught the flicker of a smile. "The cobblestones won't thank you for the courtesy."

She laughed despite herself—a short, surprised sound.

Her heart was still hammering, though she was no longer certain it was from the running.

"My hat," she said, suddenly remembering. "And my parcel—"

The wrapped gift lay in the snow a few feet away, brown paper torn, violet petals peeking through. Her hat had tumbled farther still and was currently being investigated by a very large, very damp dog.

"Allow me," the man said, releasing her gently.

"You really don't have to—"

But he was already striding toward the carnage, his boots finding purchase with far more grace than hers had managed. He scooped up the parcel and examined its torn corner.

"For someone fond of flowers?" he called over his shoulder.

"For my aunt. She gets dreadful headaches. Violets help."

He returned and offered the parcel with a small bow, as though presenting something precious. "Medicinal violets. A noble errand."

"A badly executed one," Wendy muttered, tucking the damaged package into her satchel.

The dog, meanwhile, had claimed her hat as spoils. It tugged at the velvet ribbon with enthusiasm.

"Now then," the man said, crouching to the dog's level. "My good fellow, I'll need that back. Ladies' hats don't agree with your digestion." He looked at the collar. "Ah, Nana. Well, off you go then, my lady."

He wrestled the hat free with surprising gentleness, earning a streak of dog saliva across the brim. When he stood and offered it to her, the velvet ribbon drooped with wounded dignity.

Wendy examined it. "I shall never be respectable again."

"On the contrary. You'll be the talk of the street." His eyes crinkled. "'Did you see that young lady with the battle-scarred bonnet? Must be a heroine.'"

"A heroine of clumsiness."

"The most underrated kind."

Their eyes met just for a moment. Just long enough for something to pass between them that had nothing to do with hats or violets or the softly falling snow around them. The sort of look that meant they might remember this moment later and call it *the beginning*.

Behind them, the omnibus rattled away with an apologetic clang of its bell.

"Oh no," Wendy said, but it came out softer than she intended.

"There'll be another."

"Not soon enough. I'm meant to be at the hospital, then my aunt's, then back at the hospital, and—" She stopped herself. "I'm sorry. You don't need my entire schedule."

"I don't mind schedules," he said. "I argue about them for a living. Solicitor," he added, as though this explained everything.

"Well—solicitor in training. I spend most of my time being humiliated by men in better coats."

Despite herself, Wendy smiled.

"Willem Panning," he said, extending his hand. "Of Gray's Inn. Mostly harmless."

She took it. His grip was warm even through her glove.

"Gwendolyn Smead. Of St. Bartholomew's Hospital. I bandage the people you argue over."

His laugh was sudden and bright—the sound of something cracking open to let light through.

"Miss Smead," he said, "I believe we may be natural allies." He gestured up the street, where a small open carriage waited, the driver hunched against the cold. "I was heading toward Cavendish Square. If your aunt's on the way, and the hospital's not too far after, I'd be a poor excuse for a gentleman if I let you slide there alone."

Wendy hesitated. Her mother's voice rose in her memory. *You do not accept rides from strangers, Wendy.*

But Matron's voice rose too. *The world is hard enough without refusing kindness when it finds you.*

And besides, he didn't feel like a stranger. Something in his gaze, in the way he had caught her before she fell, felt steady. Reliable. Like a window left open on purpose.

Instead of yes, she said, "I owe you an apology. And probably new gloves. You've dirtied yours."

"I consider them improved." He held up his hands to show the mud stains. "They have a story now. I'll wear it with pride."

The snow was falling harder. The dog had wandered off. The city moved around them, oblivious to the small miracle occurring on the corner—two people meeting, two lives beginning to tilt toward each other like plants toward light.

"All right," Wendy said. "Thank you, Mr. Panning."

"Willem," he corrected. "Mr. Panning is my father, and he's considerably less charming."

She took his offered arm, surprised by how natural it felt. "Willem, then. My aunt calls me Wendy."

"Wendy," he repeated.

They walked together toward the waiting carriage, their footprints filling with snow behind them.

The city looked different from the carriage seat. They rattled along through streets softened by white, past shop windows glowing with candles and Christmas displays. Wendy caught glimpses of toy soldiers in tidy rows, dolls with porcelain cheeks, gingerbread men lined up like a small, edible army.

In one window, a grandfather clock stood among the toys, its brass pendulum swinging in the dusty light.

Tick. Tick. Tick.

Counting time that children never thought to measure.

"You must see difficult things at the hospital," Willem said after a while, pulling his coat tighter against the cold.

"I see things I wish I didn't," she admitted. "And things I'm grateful for."

"That sounds like the beginning of a proverb."

"Mostly it's children with coughs and old men with lungs full of London." She paused. "Occasionally something worse."

Her gaze drifted to the streets—the gaunt faces, the huddled figures in doorways. Even in the snow, even at Christmas, the city couldn't hide its cruelties.

"There are too many lost ones," she said quietly. "Too many we don't catch in time."

Willem followed her gaze. "Every morning on my way to chambers," he said, "I pass a boy who sleeps beneath the newspaper cart. Same boy, same spot. I keep thinking I'll do something more than leave him bread. And then I'm late, or there's a deadline, or I tell myself it isn't my place."

His jaw tightened, a sign of the frustration beneath.

"Perhaps it could be," Wendy said. "Your place."

He looked at her. Something shifted in his expression—respect, recognition, the beginning of belief.

"You speak like a sermon and a revolution at once, Miss Smead."

"I speak like someone who empties bedpans for a living. And has more opinions than she probably ought," she said.

"I like opinions. They're what law is made of." He smiled. "Well, opinions and paper and the occasional judge in a truly unfortunate wig."

She laughed, and the carriage seemed less cold.

When they reached her aunt's street—the house with the ivy climbing its face like a slow green tide—Wendy gathered her satchel and examined the ruined parcel with a sigh.

"She'll have to rely on her usual tonics," she said. "The violets are beyond saving."

"We could claim they were damaged in a heroic rescue," Willem offered. "You saved a child from a runaway cart. The violets took the worst of it."

"You want me to lie to my aunt."

His eyes glinted. "I want you to tell her a story. There's a difference."

Wendy shook her head, but she was smiling.

"Thank you for the rescue, Mr.—Willem. I hope your gloves recover."

"They'll dine out on this story for years." He tipped his hat. "Safe travels, Miss Smead. And—"

He hesitated.

"Yes?"

"If you should ever find yourself late again, and in need of a carriage, I'm usually at Gray's Inn by nine. In case you want to collide with someone familiar."

Her cheeks warmed despite the cold. "I'll keep that in mind."

She stepped down from the carriage, ruined parcel in hand, hat askew, heart doing something strange and hopeful in her chest.

She did not look back as she walked to her aunt's door.

But she was smiling the entire way.

Later, he would ask her to marry him on the same street, in the same snow, on the same corner where she had nearly fallen.

She would say yes before he finished the question.

And the window—the nursery window that would come to mean everything—was already waiting for them, in a house neither of them had seen yet, in a life neither of them could imagine.

But that is a story for the next chapter.

Chapter Two

The Nursery Window

London, 1897–1901

They were married in March, when the city remembered it could bloom.

The ceremony was small, held at a registry office in front of a handful of witnesses, Wendy's mother weeping into a handkerchief she claimed was for allergies. Willem's father did not attend, having declared the match "impractical," and "fast," which Willem took as the highest possible endorsement.

Afterward, they walked through Hyde Park in a light rain, sharing an umbrella that was too small for two people but exactly right for two people who wanted an excuse to press close.

"Mrs. Panning," Willem said, testing the words.

"Mr. Panning," she returned.

"That's my father."

"Not anymore." She squeezed his arm. "You've been promoted."

He laughed—that bright, cracking-open sound she had already learned to love—and pulled her closer beneath the umbrella.

"I should warn you," he said. "I'm told I talk in my sleep. Legal arguments, mostly. Very dull."

"I'm told I steal blankets."

"Then we're perfectly matched. I'll bore you unconscious, and you'll leave me to freeze. Marriage."

She laughed and leaned her head against his shoulder.

The rain fell soft around them. The world smelled of wet earth and new leaves and the particular sweetness of a life just beginning.

***

The house on Wimpole Street was narrow and tall, wedged between its neighbors like a book pressed tight on a shelf. It had been Willem's grandmother's, then his uncle's, then nobody's for a while, accumulating dust and silence in equal measure.

But it had good bones, Willem said. And a south-facing window in the room that would become the nursery.

"Nursery," Wendy repeated, standing in the empty room. Afternoon light pooled on the floorboards. "You're very confident."

He came to stand beside her at the window. "I'm very hopeful. There's a difference."

Below, the street hummed with its ordinary music—carriages, voices, the clip of horses on stone. Above, the sky was the pale blue of early spring, streaked with clouds like brushstrokes.

"It's a good window," Wendy said.

"South-facing. Catches the morning light."

"That's not what I mean." She pressed her palm to the glass. "It feels... open. Like it's waiting for something."

Willem watched her for a moment.

"For someone," he said quietly.

She turned to look at him. His face held an expression she was learning to recognize—it meant he was seeing something she hadn't said aloud.

"You'll be a wonderful mother. When it's time," he said.

"How do you know?"

"Because you already mother everyone." He smiled. "The patients at the hospital. The boy beneath the newspaper cart—yes, I know you've been bringing him soup. Me, when I forget to eat because I'm buried in briefs."

"You do forget to eat."

"I have you to remember for me." He took her hand. "And someday, there'll be someone else. Someone small. Someone who needs this window."

Wendy looked back at the light falling through the glass. She did not know yet what the window would come to mean. How she would stand here in joy and in grief, in hope and in despair. How it would become the axis around which her entire life would turn.

She knew only that it felt right.

"Let's keep it open," she said. "When we can. I like the air."

Willem nodded.

And so the window began its long habit of openness, years before it would become ritual, before it would become prayer.

***

Peter arrived in late autumn of 1897, on a night when the wind rattled the windows and the city seemed to hold its breath.

The labor was long. Willem paced the hallway until the floorboards threatened to file a complaint. Wendy, between contractions, ordered him to sit down, eat something, stop wearing a trench in the wood. But she was glad he was near. She could hear his footsteps through the door, steady as a heartbeat.

And then, just before dawn, a cry.

Small and fierce and utterly new.

The midwife placed the baby in Wendy's arms, and the world rearranged itself around his weight. He was red-faced and furious, fists clenched, eyes squeezed shut against the indignity of being born. A shock of dark hair stood up from his head like he'd been startled. He looked, Wendy thought, like a very small, very angry philosopher.

"Peter," she said. The name she and Willem had chosen weeks ago. "Peter Panning."

The baby's eyes opened. They were unfocused, searching, finding nothing and everything at once.

Willem appeared in the doorway, pale and rumpled.

"Is he—"

"Come meet your son."

He crossed the room in three strides and knelt beside the bed. His hand trembled as he reached out to touch Peter's cheek—just one finger, barely grazing the skin, as though the baby might shatter.

"He's so small," Willem whispered.

"He'll grow."

"He's so *angry*."

"He's been evicted from somewhere warm. I'd be angry too."

Willem gave a wet, shaky laugh. His eyes were bright.

"Hello, Peter," he said. "I'm your father. I'm going to make a mess of this, probably. But I promise to try very hard."

Peter's fist uncurled. His fingers wrapped around Willem's thumb and held on.

No one spoke.

Outside, the wind gentled. The first light of morning crept through the nursery window and fell across the bed in a pale gold wash.

Wendy watched her husband and her son—this impossible, ordinary miracle—and felt something settle into place inside her chest. There was a new room in her heart, one she hadn't known was waiting to be built.

"The window," she said softly.

Willem looked up. "What?"

"Open it. Just a little. I want him to hear the world."

Willem crossed to the window and pushed it open an inch. Cold air slipped through, carrying the smell of autumn leaves and chimney smoke and the sound of London waking.

Peter turned his head toward the draft. His eyes, still unfocused, seemed to search for something beyond the glass.

"He's listening," Willem said.

"He's always going to be listening." Wendy didn't know why she was so certain. "He's that kind of child."

Willem returned to the bed and gathered them both in his arms, the whole of his world contained in a single embrace.

"Thank you," he murmured into her hair.

"For what?"

"For this. For him. For the collision."

She smiled, exhausted and radiant.

"You're welcome. I'll try to knock you down again sometime."

He kissed her forehead.

And Peter, between them, slept at last—fingers curled around his father's thumb, chest rising and falling with the rhythm of new life.

***

The months that followed were a blur of sleepless nights and small victories.

Peter's first smile. His first laugh—a startled, hiccupping sound that made Willem drop a stack of legal papers and declare it the finest closing argument he'd ever heard.

His first tooth. His first word ("no," naturally). His first steps, taken toward the nursery window with arms outstretched, as though he meant to climb through and explore whatever lay beyond.

"He's drawn to it," Willem observed one evening, watching Peter press his small hands against the glass.

"He likes the light."

"He likes the *outside*." Willem lifted Peter onto his hip. "What do you see out there, little man?"

Peter pointed at something—a bird, a cloud, a passing carriage—and babbled with great seriousness.

"He's giving a legal opinion," Willem said. "I recognize the tone."

Wendy laughed and joined them at the window.

The city spread before them in the evening light—rooftops and chimneys, church spires and distant parks. Ordinary and miraculous, as cities are.

"We should hang something," she said. "In the window. Something that catches the light."

"A prism?"

She considered it. "Yes. I had one as a child. I loved the way it scattered rainbows."

"Then we'll find one."

They did—a small glass prism on a silver chain, purchased from a shop near Covent Garden. Wendy hung it in the nursery window, where it caught the afternoon sun and threw colors across the walls.

Peter was enchanted.

He would sit for long stretches, watching the rainbows move across the floor. Sometimes he chased them, laughing when they slipped away. Sometimes he simply watched, quiet and intent, as though the shifting light was telling him a story only he could hear.

"Where does the light go?" he asked one day, when he was old enough for questions. "When it leaves the prism?"

Wendy knelt beside him. "It goes everywhere, into everything. Light doesn't disappear, darling. It just finds new places to land."

Peter considered this with the gravity of a scholar. "Even in the dark?"

"Especially in the dark. That's when it matters most."

He nodded slowly, then reached up and touched the prism, setting it swinging. Colors scattered across his face—red, gold, violet, green.

"I want to go where the light goes," he said.

Wendy's heart clenched with a fear she did not yet understand.

"Someday," she said, pulling him close. "But not yet. Not for a long time."

***

At night, she told him stories.

This was their ritual: bath, milk, the rocking chair by the window. Peter would curl against her chest, thumb in his mouth, eyes heavy with sleep but fighting it—always fighting it—because he wanted one more story, just one more.

She told him fairy tales at first. The old ones, with their wolves and witches and children lost in dark woods. But Peter didn't like those. The lost children frightened him.

So she invented new stories. Gentler ones. Stories about boys who found their way home, about mothers who searched and never gave up, about windows that stayed open no matter how cold the night was.

"Tell me about the boy who could fly," Peter murmured one evening.

Wendy paused. She didn't remember telling that story.

"What boy?"

"The one who flies out the window." Peter's eyes were half-closed, his voice drowsy. "And finds the place where children don't get lost."

She stroked his hair, unsettled by something she couldn't name.

"Where did you hear about that?"

"I dreamed it."

The window was open a crack. Cold air seeped through, stirring the curtains.

"Tell me," Peter said. "Tell me about Neverland."

The word hung in the air. Wendy was certain she had never spoken it. But Peter was waiting, and he was warm against her chest, and outside the window the stars were appearing one by one like small steady promises.

So she told him.

She told him about an island where children never grew old, where the trees whispered secrets and the stars hung close enough to touch. She told him about mermaids and fairies and a tribe of lost boys who had found each other in the dark.

She told him about a boy who flew—not because he had wings, but because he believed he could. A boy who led the lost ones home.

Peter fell asleep smiling.

Wendy held him for a long time afterward, watching the prism sway in the draft from the window.

Where did he hear that word?

She didn't know.

But something about the story felt true in a way she couldn't explain. As though it had been waiting for someone to tell it—waiting for her son, with his window-watching eyes and his dreams of flight.

She kissed his forehead and carried him to his crib.

"Good night, my darling," she whispered. "Good night, my little Peter."

Outside, the wind picked up. The window rattled softly in its frame.

And in his sleep, Peter smiled, as though he could hear something calling to him from very far away.

***

She would remember this later. The stories. The window. The way he listened to things she couldn't hear.

She would wonder if she had somehow summoned it—the darkness that came for him. If the stories had been a map, and she had drawn it without knowing.

But that was guilt speaking, not truth.

The truth was simpler and more terrible:

Some children are taken.

Some windows are opened.

And some mothers spend the rest of their lives learning that love is not enough to keep anyone safe. But it is enough to keep searching.

It is always, always enough to keep searching.

Chapter Three

The Child Stolen

London, Winter 1901

Later, Wendy would try to remember if there had been signs.

A cold spot in the nursery that hadn't been there before. A shadow that moved against the light. The way Peter had been restless that week, waking in the night to stand at the window, pressing his palms to the glass as though answering a summons only he could hear.

"What do you see?" she had asked him, lifting him back into bed.

"The stars are talking," he'd said.

"What are they saying?"

He had looked at her with those strange, searching eyes—her son and also something else, something she couldn't name—and said: "They're saying *come*."

She had tucked the blankets around him and kissed his forehead and told herself it was only a dream.

Children dreamed.

Children imagined.

It didn't mean anything.

***

The night it happened was cold—bitter cold, the kind that crept beneath doors and through the seams of windows, the kind that made the city pull its collar up and hurry home.

Wendy checked on Peter at nine o'clock. He was asleep, one arm flung over his head, the other clutching the stuffed rabbit Willem had given him. The window was latched. The prism hung still in the darkness. The candle on the bedside table burned low but steady.

Everything was ordinary.

Everything was safe.

She went to bed beside Willem, who was reading a legal brief and losing the battle against sleep. She kissed his cheek. He murmured something about tort law. She smiled and closed her eyes.

The clock in the hallway struck midnight.

Then one.

Then two.

At half past three, Wendy woke.

She didn't know why. No sound had disturbed her, no movement, no cry. But something had shifted in the air. It was a change in pressure, a held breath, a particular silence that comes when something has gone terribly wrong.

She sat up.

Willem slept on beside her, oblivious.

The house was quiet. Too quiet. The kind of quiet that pressed against the ears.

Wendy rose. Her feet found the cold floor. She didn't bother with slippers or a robe—she just walked, pulled by something she couldn't name, down the hallway toward the nursery.

The door was open.

She had closed it. She always closed it.

Cold air poured through like water through a breached hull as she stepped inside.

The window not just unlatched but flung wide, curtains billowing in a wind that shouldn't have been strong enough to move them. Snow swirled in, already dusting the floorboards, the windowsill, the empty bed.

The empty bed.

The blankets were thrown back. The stuffed rabbit lay on the floor, as though dropped in haste. The pillow still held the impression of a small head.

But Peter was gone.

***

Wendy did not scream.

Later, this would surprise her. She had always imagined that if something happened to her child—the unthinkable, the unspeakable—she would scream until her throat tore, until the world cracked open, until someone came to tell her it wasn't true.

Instead, she stood very still. The cold wrapped around her like a second skin. Snowflakes landed on her face and did not melt.

She looked at the window.

She looked at the bed.

She looked at the floor, searching for—what? Footprints? Signs of a struggle? Some explanation that would make this make sense?

There was nothing. Only snow. Only silence. Only the curtains moving in a wind that seemed to come from somewhere other than outside.

"Peter," she said. Her voice was calm. Conversational. As though he might be hiding, as though this were a game. "Peter, darling, where are you?"

No answer.

"Peter."

The wind gusted. The prism swung on its chain, scattering no rainbows—there was no light to catch—just moving, back and forth, like a pendulum counting down.

"PETER!"

The scream came at last. It tore out of her chest and filled the room, filled the house, filled the street below where a few startled pedestrians looked up at the sound of a mother's heart breaking.

Willem appeared in the doorway, wild-eyed, still half-asleep. "Wendy? What is it? What's—"

He saw the empty bed. The open window. His wife standing barefoot in the snow that had blown in, her nightgown soaked, her face the color of ash.

"He's gone," she said. "Willem. He's gone."

***

The police asked their questions with practiced patience. *When did you last see the child? Was the window locked? Had anyone suspicious been seen in the neighborhood? Did the boy have a habit of sleepwalking?*

Wendy answered as best she could, her voice steady even as her hands shook. Willem paced behind her, running his fingers through his hair until it stood in wild peaks.

The officers exchanged glances. She saw them do it—the silent communication of men who had seen cases like this before, who knew how they usually ended.

"We'll do everything we can, Mrs. Panning," the inspector said. His voice was kind. She hated it. "Children wander off sometimes. Get confused in the cold. We'll organize a search."

Wendy's voice was flat. "He didn't wander off. He was taken."

"Is there any reason to believe—"

"The window was *opened*. From the outside. He didn't open it himself. Someone came *in*."

The inspector wrote something in his notebook. She couldn't see what.

"We'll look into all possibilities," he said. "In the meantime, if you could provide a recent photograph, a description of what he was wearing..."

The words blurred. Wendy heard herself answering—*dark hair, brown eyes, four years old, small for his age, wearing a white nightshirt with blue stitching*—but it felt like listening to someone else speak. Someone far away. Someone who still believed this could be fixed.

The police left. The sky lightened toward a dawn that seemed obscene—how dare the sun rise when her child was gone?—and

Wendy stood at the nursery window, the open window, and did not move.

***

They searched for weeks.

Willem hired private investigators. Wendy walked the streets herself, clutching a photograph, stopping strangers, describing Peter to anyone who would listen. She visited hospitals, orphanages, workhouses. She checked the river, though she couldn't say why. She checked the parks where children played, scanning every small dark head for the one that belonged to her.

Nothing.

No sign. No sighting. No ransom note. No body.

He had simply vanished, as though the night had swallowed him whole.

"Someone took him," Wendy said, again and again. "Someone came through that window."

But who had climbed through a second-story window in the middle of winter? Who had taken a child and left no trace? The investigators had no answers. The police had no answers. The neighbors whispered about unlocked windows and negligent mothers and the tragedy of it all.

Wendy stopped leaving the house. She stopped eating, stopped sleeping, stopped doing anything except sitting in the nursery, in the rocking chair by the window, watching and waiting.

"Gwen." Willem knelt beside her one evening, his face gaunt with grief. "Gwen, you have to rest. You'll make yourself ill."

"I can't leave the window."

"Why?"

"Because—" Her voice cracked. "Because what if he comes back? What if he finds his way home, and the window is closed, and he thinks we've given up?"

"We won't give up. We'll never give up. But you sitting here, not eating, not sleeping... it won't bring him back."

She gripped the arms of the chair. "You don't know that. You don't know anything. No one knows anything."

Willem's eyes were wet. He looked older than he had a month ago—much older, as though grief were aging him in dog years.

"What if—" He stopped. Started again. "The inspector said—"

"I know what the inspector said."

"It's been three weeks. If someone had taken him for ransom, we would have heard by now. And if it was something else—"

"Don't."

"Wendy—"

"*Don't.*" She turned to face him. Her eyes were terrible, fierce and hollow at once, like a fire burning in an empty room. "He's alive. I would know if he wasn't. I would *feel* it. He's alive, and he's somewhere, and he needs me to keep watching. He needs me to keep the window open."

Willem stared at her for a long moment. Then he leaned forward and pressed his forehead to hers.

"All right," he whispered. "All right. We keep watching. We keep the window open."

"Promise me."

"I promise."

"No matter how long."

"No matter how long."

She let out a breath she didn't know she'd been holding.

Outside, snow began to fall again—soft, relentless, covering the city in white like a page wiped clean.

***

Winter turned to spring. Spring to summer. Summer to autumn.

No Peter.

Wendy returned to work at the hospital—not because she had healed, but because she could not sit still any longer. The waiting was worse than the doing. At least in the wards, she had hands to hold and bandages to change and children who needed her.

Other people's children.

She threw herself into their care with a ferocity that startled her colleagues. She sat with the dying ones when no one else would. She learned their names—not just the names on their charts, but the names they called themselves, the nicknames their mothers used, the secret names they whispered in their sleep.

She wrote the names in a small notebook she kept in her apron pocket.

"Why do you do that?" a young nurse asked her once.

"Because someone should remember them," Wendy said. "Even if no one else does."

The nurse looked at her with something like pity.

Wendy didn't care. She kept writing names.

At night, she returned to the nursery. She sat in the rocking chair. She watched the window.

Willem joined her sometimes. They didn't speak—there was nothing left to say—but they sat together in the dark, two people drowning in the same grief, holding on to each other because there was nothing else to hold.

The prism gathered dust on its chain, but Wendy couldn't bring herself to clean it. Couldn't bring herself to touch anything in the room, as though preserving it exactly as it had been might somehow preserve the possibility of Peter's return.

Come back, she thought, staring at the stars beyond the glass. *Please come back. The window's open. I'm waiting.*

The stars didn't answer.

But she kept waiting anyway.

***

One year passed.

The investigators gave up. The police closed the case. The neighbors stopped asking, stopped bringing casseroles, stopped looking at her with that particular mixture of pity and discomfort that marked her as a woman who had lost something that couldn't be recovered.

Wendy had become someone else. Someone quieter. Someone with a stillness at her center that hadn't been there before, like a pond that had frozen solid and learned to call it peace.

She still worked at the hospital. Still wrote names in her notebook. Still sat at the window every night, even though Willem had stopped joining her, even though the house had grown silent around their grief.

She did not give up.

She did not close the window.

And one night in autumn, nearly one year after Peter had vanished, she was sitting in the rocking chair when she heard a sound.

A tap.

Soft. Hesitant. Like a bird striking glass.

Then another.

Tap. Tap.

Wendy's heart stopped.

She rose slowly, not daring to breathe, not daring to hope.

And there—crouched on the narrow ledge outside, shivering in the cold, wearing clothes she didn't recognize and a face she would have known anywhere—

"Peter?"

The boy looked up at her. His eyes were wild, dark-ringed, full of something she couldn't name. His hair was longer, matted with dirt and leaves. His hands were pressed against the glass, and they were shaking.

"Mama?" His voice was hoarse, cracked, as though he hadn't used it in a long time. "Mummy, I found the window. I came back."

Wendy didn't remember pulling him through. Didn't remember falling to her knees with her arms wrapped around him so tightly he gasped for breath. She only remembered the smell of him—different now, tinged with earth and something else, something cold and green and far away—and the sound of her own voice, broken and whole at once, saying his name over and over like an incantation.

"Peter. Peter. Peter. My Peter. My boy. My darling boy."

He clung to her, trembling.

"I came back," he said again. "I didn't forget. I came back."

"I know. I know. I kept the window open. I never stopped waiting," she whispered.

He pulled back to look at her. His eyes—those strange, searching eyes—were the same, but also different. Older. Deeper. As though he had seen things no child should.

"They wanted me to stay," he said. "But I remembered you. I remembered the window. I followed the light home."

Wendy didn't ask who *they* were. Didn't ask where he had been. Not yet. Not tonight.

Tonight, she only held him.

Willem's footsteps thundered up the stairs. He burst through the door and froze, mouth open, eyes wide, staring at the impossible.

"Peter?"

"Papa."

Willem crossed the room in two strides and fell to his knees beside them. He gathered them both in his arms—wife and son, the whole of his shattered world suddenly, impossibly, made whole.

They stayed that way until dawn.

And when the sun rose, its light caught the prism in the window—the dusty, forgotten prism—and scattered rainbows across the room.

Peter watched them move across the walls.

"I told you," Wendy whispered, remembering a conversation from another life. "Light doesn't disappear. It just finds new places to land."

Peter nodded slowly. But his eyes, she noticed, didn't follow the rainbows.

They followed the shadows.

Chapter Four

The Child Returned

London, 1902

Peter slept for three days.

Not peacefully—never peacefully—but deeply, as though his body was trying to make up for months of vigilance. He curled in his bed with his knees drawn to his chest, face pressed into the pillow, one hand always reaching toward the window.

Wendy sat beside him, not leaving except when Willem forced her to seek rest she couldn't find. Otherwise, she watched her son's chest rise and fall, counting his breaths like prayers.

He was thinner than before. She could see the knobs of his spine through his nightshirt, the sharp angles of his shoulder blades. His feet were calloused, toughened, as though he'd walked a long way without shoes. There was dirt beneath his fingernails that wouldn't come clean, and scratches on his arms that looked like they'd been made by thorns.

The doctor pronounced him malnourished but otherwise healthy.

"Remarkable, really," the doctor said, packing his bag. "Children are resilient. A few weeks of good food and rest, and he'll be right as rain."

Wendy nodded. Thanked him. Showed him to the door.

She didn't believe a word of it.

***

On the fourth day, Peter woke.

Wendy was in the rocking chair, drifting in and out of an exhausted half-sleep, when she heard the rustle of sheets. Her eyes flew open.

Peter was sitting up in bed, staring at her with an expression she couldn't read.

"You stayed," he said.

"Of course I stayed. I'll always stay."

"They said you wouldn't. They said mothers forget. They said the window would be closed." His voice was strange—older than it should have been, carrying the cadence of someone who had learned to speak in darkness. "But I knew. I knew you'd be waiting."

Wendy moved to the bed and gathered him into her arms. He stiffened at first—a flinch, automatic and heartbreaking—then slowly, slowly, relaxed against her.

"I will always wait for you," she said into his hair. "Always. There is no place you could go, no length of time you could be away, that would make me stop waiting."

"Even if it takes years?"

"Even if it takes forever."

He was quiet for a long moment. Then, very softly, he said, "It felt like forever. Where I was."

Wendy's arms tightened. "Where were you, darling? Can you tell me?"

Peter pulled back to look at her. His deep, strange eyes held something vast and terrible and not entirely childlike.

"I don't have the words," he said. "Not in this language. Not in any language that belongs to the daylight." He looked toward the window, still open, still framing the sky. "There was a place. An island. It was beautiful, sometimes. And terrible, sometimes. And there were others—other children who'd been taken, who'd forgotten how to go home."

"But you remembered."

"I remembered you." He touched her face, tracing the lines that grief had carved around her eyes. "You were the light in the window. Every time I started to forget, I would see you. Not really—not with my eyes—but here." He pressed his hand to his chest. "Here, I could see you waiting. And I knew I had to come back."

"And now you're home."

"Now I'm home." But his voice held something uncertain. Something not quite complete. "Most of me, anyway."

"What do you mean?"

Peter didn't answer. He was looking at the floor, at the place where his shadow fell across the sunlit boards.

"It's still attached. I wasn't sure it would be," he said quietly.

Wendy didn't know what to say. So she simply held him, watching the shadows move across the room and wondering what kind of darkness had taught her son to doubt his own wholeness.

***

The weeks that followed were a slow, careful reconstruction. Peter ate when food was placed before him, but he never asked for it. He slept when exhaustion took him, but he fought it as long as he could, jolting awake at the slightest sound. He spoke when spoken to, but he didn't volunteer words. It was as though he had forgotten how to want things—or had learned, in that other place, that wanting was dangerous.

"Give him time," Willem said. "He's been through something we can't understand. He needs time to find his way back."

"What if he can't?" Wendy's voice broke on the question. "What if part of him is still there? What if he left something behind that he can't get back?"

"Then we love what's here. We love him exactly as he is, and we trust that love is enough."

"Is it? Is love enough?"

Willem pulled her close. "It's what we have. It has to be."

***

The obsession with shadows began in summer.

Wendy noticed it first in the garden. Peter was crouching on the grass, utterly still, watching his own shadow stretch across the lawn in the afternoon sun.

"What are you doing, darling?"

"Watching."

"Watching what?"

"To see if it moves on its own."

She knelt beside him, the grass cool beneath her knees. "Shadows don't move on their own. They follow us."

He looked up at her with an expression so serious it made her chest ache. "Are you sure? What if some shadows get tired of following? What if they want to go ahead instead?"

"Why would a shadow want to go ahead?"

"To warn you. Or to lead you somewhere." He returned his gaze to the dark shape on the grass. "Or to run away."

"Has your shadow ever run away, Peter?"

A long pause. The garden was quiet around them, bees humming in the roses, birds calling from the hedge.

"Not yet," he said finally. "But I think it wants to."

***

She told Willem about the conversation that night, after Peter was asleep.

"Shadows that run away," Willem repeated. He was sitting in his study, a glass of whiskey untouched beside him. He'd started pouring drinks in the evenings and then forgetting to drink them, as though the gesture of normalcy was more important than the thing itself. "What does that mean?"

"I don't know. He's been watching shadows constantly. His own, mine, the ones in corners. He talks to them sometimes."

"Talks to them?"

"Whispers. I can't hear what he says."

Willem rubbed his face with both hands. "Should we call a doctor? A different doctor? Someone who specializes in..." He gestured vaguely. "Children who've been through things?"

"And tell them what? That our son was kidnapped by something that came through his window, taken somewhere he can't name, and returned speaking of shadows that want to escape? They'll think we're mad. Or that he is." Wendy's voice was sharper than she intended.

"Maybe he is."

"Don't."

"Wendy—"

"He's not mad." She stood, arms wrapped around herself. "He's *wounded*. There's a difference. Wounds can heal."

"Some wounds don't heal. Some just become part of you."

She looked at her husband—at the gray in his hair that hadn't been there a year ago, at the lines around his eyes, at the way he held himself like something was broken inside.

"Then we help him carry it," she said. "That's all we can do. We help him carry it until he's strong enough to carry it himself."

Willem was quiet for a long moment. "What if he's never strong enough?"

"Then we carry it forever."

She meant it. She would spend the rest of her life proving it.

***

Peter spoke of shadows again on an autumn afternoon, when the light was golden and slanting and the world smelled like dying leaves.

He was sitting by the nursery window—unlatched now, open a crack—watching the shadows of branches move across the floor. Wendy sat in the rocking chair, mending a torn shirt, keeping him company the way she'd learned to do. Present but not pressing. Available but not demanding.

"Mama?"

"Yes, darling?"

"What happens if your shadow gets lost? Would you still be whole?"

"I think so. A shadow is just... an absence of light. It's not part of you, exactly. It's just a shape you make."

"But what if it wasn't?" He looked up at her. "What if they were the parts of us that go to dark places, so we don't have to?"

Wendy's breath caught. "Peter—"

"When I was... away..." He stopped. Started again. "There were places without light. Not just dark, but *no* light. Ever. And I thought—I thought maybe my shadow was there, in those places, feeling the dark so I didn't have to."

"And now?"

"Now I'm home." His voice dropped to a whisper. "But sometimes I think my shadow is still there. Still in the dark places. Waiting for me."

Wendy moved to sit beside him on the floor. She didn't reach for him—she'd learned to wait for permission—but she sat close enough that he could lean into her if he wanted.

"If your shadow were ever lost," she said quietly, "I would sew it back on."

Peter looked at her. "You can't sew a shadow."

"I could try. I would try anything, Peter. I would travel to whatever dark place took it and bring it back and stitch it to your heels myself." She met his eyes steadily. "That's what mothers do. We find the lost things. We make them whole again."

He stared at her for a long moment. Then, slowly, he leaned into her side—the first time he'd initiated contact since his return. She wrapped her arm around him and held her breath, afraid any movement might startle him away.

"The other children are still there," he said quietly. "They couldn't find the window. They forgot how to look."

"But you remembered."

He turned to look at her, and his eyes were bright with tears he wouldn't let fall. "I remembered you telling me about the light. How it scatters but doesn't disappear. I followed it home."

Wendy gathered him fully into her arms. He stiffened at first—he always did—then slowly, slowly, relaxed against her.

"That's why the window stays open," she murmured into his hair. "For you. For all the children who need a way home. I'll keep it open forever if I have to."

Peter pulled back to look at her.

"What if the bad things come through too?"

"Then we'll face them together."

He considered this. Then, slowly, he nodded.

"Together," he repeated, as though testing the word.

It was, Wendy thought, the first time since his return that he'd sounded anything like himself.

***

And then there was Charlie.

Peter found him in the alley behind the baker's shop on a Tuesday in late November, huddled between rubbish bins with his arms wrapped around his knees and a bruise blooming across his cheekbone.

"What are you doing?" Peter asked.

The boy looked up. He was about Peter's age, with red hair and sharp eyes and the particular stillness of a child who had learned not to draw attention.

"Waiting," he said.

"For what?"

"For my father to stop being angry."

Peter considered this. He sat down on an overturned crate, heedless of his clean clothes, and studied the boy with the same intensity he usually reserved for shadows.

"Does he get angry often?"

"When he drinks. Which is always."

"Does he hit you?"

The boy's jaw tightened. He didn't answer. He didn't have to.

"Come home with me," Peter said.

"What?"

"Come home with me. My mum will feed you. She feeds everyone." Peter stood and extended his hand. "I'm Peter."

The boy stared at the offered hand like it might bite him. But something in Peter's face—some certainty, some unshakeable calm—seemed to reassure him.

"Charlie," he said, taking the hand. "Charlie Morton."

Peter pulled him to his feet. "Well, Charlie Morton. Let's get you some supper."

***

Wendy found them in the nursery an hour later. Peter sat cross-legged on the floor, demonstrating how the prism scattered light while Charlie watched with wary fascination.

"And see, when you move it, the rainbows move too. They're not stuck. They go wherever the light takes them."

"Peter," Wendy said gently. "Who's your friend?"

"This is Charlie. He was hiding behind the baker's. His father hits him." Peter said it matter-of-factly, the way another child might announce that their friend had a dog. "He needs supper."

Wendy looked at the smaller boy. He was thin, shivering, and eyeing her with the wariness of a child who'd learned not to trust adults.

"Of course," she said. "Charlie, you're very welcome here. Supper will be ready in half an hour."

Charlie ate three helpings of everything. Peter watched him with fierce attention, making sure his plate was never empty, fetching him more bread, more milk, more pudding.

After supper, Wendy walked Charlie home. His building was in a poorer part of the neighborhood, the stairs narrow and creaking, the walls thin enough to hear arguments bleeding through from every floor.

Charlie's mother answered the door. She was small—smaller than Wendy had expected—with red hair like her son's and the worn-down look of a woman who had been fighting a losing battle for years. Her eyes went immediately to Charlie, then to Wendy, and something like shame crossed her face.

"Thank you," she said. "He wanders off sometimes. Gets into trouble."

Wendy kept her voice gentle. "He's welcome at our house anytime. I'm Wendy Panning. We live on Wimpole Street. Number fourteen."

"Lilly Ann Morton." The woman's hand was rough when Wendy shook it, her grip uncertain. "Thank you. For feeding him. You didn't have to—"

"I wanted to. Peter wanted to. Charlie's a lovely boy."

Lilly Ann's eyes flickered—something between pride and pain. "He is. He's the best thing I've ever done." She glanced over her shoulder, toward the darkened flat behind her. "I try to keep him safe. I try to—"

"I know." Wendy squeezed her hand. "I know. Bring him to us whenever you need to. Day or night."

Lilly Ann stared at her—really stared, as though Wendy had said something in a language she'd almost forgotten.

"Thank you," she whispered. "Thank you, Mrs. Panning."

"Wendy. Please."

***

When she returned home, Peter was waiting in the nursery, sitting by the window.

"I want to help them." He turned to look at her, and his eyes held that strange, deep certainty that had come back with him from wherever he'd been. "The lost ones. The ones who don't have a window. I want to help them find their way home."

Wendy's heart swelled and broke at the same time.

"You will, darling," she said. "When you're older. When you're ready."

"How will I know when I'm ready?"

She had no answer. So she simply pulled him close and held him, and watched the first stars appear in the darkening sky beyond the window.

She didn't know, then, what shape his helping would take.

She didn't know about the war that was coming.

She didn't know that her son would carry his need to save lost boys all the way to the trenches of France, running toward thunder because running away was something he'd never learned how to do.

She only knew that he was home. That he was hers. That for now, at least, the window was open, and the light was scattering rainbows across the walls, and Peter was watching—not the light, not the shadows, but somewhere in between.

It was enough. It had to be enough.

Chapter Five

The Storyteller

London, 1903

Wendy met James Barrie on a Tuesday, in a hospital corridor that smelled of carbolic soap and sorrow.

She had been volunteering at Great Ormond Street since Peter's return. It was a way to fill the hours, to do something useful, to practice the kind of watching that had kept her sane through the years of waiting. She read to the children, wrote letters for the ones too weak to hold a pen, sat with the ones who had no one else to sit with them.

It was there, in the children's ward, that she learned to see the different kinds of lost.

There were children lost to illness, their bodies failing them in ways their minds couldn't understand. Children lost to poverty, whose parents had surrendered them to the hospital because they couldn't afford to keep them. Children lost to violence, whose bruises told stories their voices couldn't speak.

And then there were the ones like Peter had been—lost to something stranger, something that didn't have a name, something that had taken them somewhere and brought them back *different*.

She was writing a letter for a boy named Thomas. He was twelve years old, dying of something the doctors couldn't cure, dictating his final words to a mother who would never read them in time, when she noticed the small man watching from the doorway.

He was slight and neat, with large, sad eyes that seemed to see everything and miss nothing. His clothes were expensive but worn carelessly, as though fashion was a language he spoke fluently but didn't care about. He stood utterly still, watching her with the children, and there was something in his gaze that made her think of Peter.

Something *hungry*.

"Can I help you?" she asked.

"I don't know. Can you?" His accent was Scottish, his voice soft, with an undertone of something that might have been humor or might have been heartbreak. "I've been watching you with them. The children. You're remarkable."

"I'm their friend. Nothing remarkable about that."

"Oh, but there is. Most people can't bear to look at dying children. They turn away. But you look *at* them. You see them. There's a difference."

Wendy set her pen down. Thomas had fallen asleep, his letter unfinished, his hand still curled around hers.

"Who are you?"

"James Barrie." He stepped into the room, moving carefully, as though afraid to disturb something fragile. "I'm a playwright. I write stories for children. Or about children. Sometimes I'm not sure which."

"I've heard of you." Wendy studied him with the same directness she'd learned to turn on Peter's shadows. "You wrote *The Little Minister*."

Something flickered in his eyes. "Mrs. Panning, tell me. Do you have children of your own?"

"One. A son. Peter."

"And where is Peter now?"

Wendy looked down at Thomas's sleeping face. At the fragile rise and fall of his breathing. At the letter that would never be finished, the goodbye that would never be said.

"Home," she said finally. "He's home now. But for a long time, he was lost."

Barrie sat down in the chair across from her. His movements were slow, deliberate, as though he was approaching a wild animal.

"Tell me," he said.

And—because she was tired, because she had never told anyone, because there was something in this strange little man that reminded her of her son—she did.

***

She told him everything. Not just the facts—she had repeated those so many times they had lost their meaning, become rote, become the *story* of Peter's disappearance rather than the thing itself. She told him the truth: the window, the prism, the light that scattered like hope. The darkness that had come through and taken her child. The years of waiting, of watching, of refusing to close the window even when everyone said she should.

And she told him about Peter's return.

"He was different," she said. "Still Peter—still my Peter—but changed. He watches shadows now. He talks about things without names, places without light. He's seen something that children shouldn't see, and it lives in him now. It always will."

Barrie was silent for a long time.

"I lost children too," he said finally. "Not the way you did. My wife and I—we never had any of our own. But I watched the Llewelyn Davies boys grow up, and I loved them as though they were mine. Michael. George. Peter." He smiled faintly. "Yes, another Peter. The world is full of them. And I've spent years trying to understand what childhood really is. What we lose when we leave it. What some children never lose at all."

"And what have you learned?"

"That the window matters." He leaned forward, his eyes bright with something fierce. "The window the mother keeps open. The light she leaves burning. That's what brings them home. Not magic, not wishes. Just love, stubborn and patient, refusing to give up."

Wendy felt tears prick at her eyes. "That's what I told Peter. When he was small. Before he was taken."

"Then you gave him the truth. The only truth that matters." Barrie reached out and touched her hand, very gently. "May I meet him? Your Peter? I'd like very much to meet the boy who found his way home."

***

She brought Barrie to Wimpole Street that Sunday.

Peter was in the nursery when they arrived, standing at the window, watching the afternoon light refract through the prism.

Charlie Morton was there too—a fixture now, appearing most days after school, eating supper with them more often than not, slowly learning to trust that this house would not hurt him.

"Peter," Wendy said. "This is Mr. Barrie. He's a playwright."

Peter turned. His eyes moved over Barrie with that particular scrutiny he had developed since his return—measuring, assessing, searching for threats in the shape of a person's shadow.

"You write stories," Peter said.

"Yes."

"What kind of stories?"

"All kinds. But mostly stories about children. About what happens to them when adults aren't looking."

Peter's voice was flat. "I know what happens. I was there."

Barrie didn't flinch. "Yes. Your mother told me. I'm very sorry for what you went through."

"Are you going to write about it?"

The question hung in the air. Wendy started to intervene, but Barrie raised a hand.

"I don't know," he said honestly. "I write about lost children all the time. But I've never met one who came back. You're the first."

"I'm not lost anymore."

"No. But you remember what it was like. Don't you? You remember the island."

Peter's face went very still. "How do you know about the island?"

"I know about *an* island. I've been dreaming about it for years. A place where children go when they're forgotten. Where time doesn't work properly. Where you can fly if you believe hard enough." Barrie's voice was soft, almost hypnotic. "I call it Neverland."

"That's not what it's called."

"What is it called?"

Peter was silent for a long moment. Charlie had drawn closer, watching the exchange with wary eyes.

"It doesn't have a name," Peter said finally. "Not in any language that belongs here. But Neverland..." He tested the word. "Neverland is close enough."

***

Barrie came back the following week. And the week after that. And the week after that.

He and Peter would sit in the nursery for hours, talking in low voices, drawing maps of places that didn't exist on any chart. Wendy watched from the doorway, unsure whether to be worried or grateful. Barrie was gentle with Peter—gentler than most adults knew how to be with a wounded child—and something in their conversations seemed to help. Peter laughed more often. He chased the rainbows again, sometimes. He slept through more nights than he didn't.

"He's giving me the story," Barrie told Wendy one evening. "The real story. Not the darkness—I won't put the darkness on a stage—but the shape of it. The truth underneath."

"What truth?"

"That children get lost. That some of them find their way home, and some of them don't. That the window matters—the mother's love, the refusal to give up, the light that never goes out."

His eyes were bright. "That's the story I want to tell. Not the tragedy. The hope."

"And Peter? What does he think about being in a story?"

"He thinks..." Barrie paused, choosing his words carefully. "He thinks that if I tell it right, other children might find their way home. He wants to help them. Even the ones he'll never meet."

Wendy looked toward the nursery, where Peter was showing Charlie how to hold the prism to catch the light at just the right angle.

"That's all he's ever wanted," she said. "To help the lost ones."

"Then let me give him this. A story that matters. A myth that might actually save someone."

***

The play premiered in December 1904.

Wendy sat in the audience with Willem on one side and Peter on the other. Peter was still and watchful, his eyes fixed on the stage with an intensity that made her heart ache.

The curtain rose on a nursery. A window, open. A girl named Wendy telling stories to her brothers. And then—

A boy.

A boy who flew through the window, searching for his shadow, searching for the stories he'd lost, searching for something he couldn't name but knew he needed.

Peter Pan.

Wendy—the real Wendy—watched her son see his story become myth.

She saw his hands grip the armrests when the shadow was sewn back on. Saw him lean forward when the children flew to Neverland. Saw him flinch, almost imperceptibly, when Captain Hook appeared—a figure of darkness, of relentless pursuit, of the thing that follows you no matter how far you fly.

And she saw him weep—silently, shoulders shaking—when Wendy Darling stood at the window and promised Peter Pan that she would leave it open. That she would always, *always* leave it open.

"To die," the boy on stage declared, "would be an awfully big adventure!"

The audience laughed. It was meant to be funny—the bravado of youth, the immortal confidence of a child who doesn't understand mortality.

But Wendy saw her son's face.

He wasn't laughing.

He was *memorizing*.

***

Afterward, in the lobby, Barrie found them.

"Well?" he asked Peter. "What did you think?"

Peter was quiet for a long moment. His face was unreadable.

"The lost boys," he finally said. "In the story. They don't have mothers."

"No. That's why they're lost."

"But Wendy takes care of them. Tells them stories. Tucks them in."

"Yes."

"So she becomes their mother. Even though she's just a girl."

Barrie nodded, waiting.

"And then she goes home." Peter's voice was flat. "She goes back through the window and grows up and forgets them."

"She doesn't forget. She—"

"She *leaves.*" Peter looked up at Barrie. His eyes were ancient in his young face. "She leaves them in Neverland, and they stay lost forever."

Barrie opened his mouth to respond, but Peter had already turned away, pushing through the crowd toward the exit.

Wendy caught Barrie's arm.

"He's not angry at you. He's angry at the truth," she said softly.

"What truth?"

"That you can't save everyone. Some lost boys stay lost." She looked toward where Peter had disappeared. "He's already planning to prove you wrong."

"How?"

"I don't know yet. But when he figures it out, God help anyone who stands in his way."

***

The years that followed were strange ones.

Peter Pan became a phenomenon. Everyone knew the story—the boy who flew, the girl who mothered, the window that never closed. Children begged their parents to leave nursery windows open. Actresses competed for the role of Peter. The phrase "to die would be an awfully big adventure" entered the language like a key into a lock.

And Peter—the real Peter—grew up in the shadow of his own myth.

He was a serious child. Too serious, Willem sometimes worried. He read voraciously—history, geography, medicine, anything that might teach him how to help, how to save, how to reach the lost and bring them home. He volunteered at the hospital with his mother, carrying trays and reading to patients and learning, always learning, how to sit with suffering without being destroyed by it.

Charlie Morton became his shadow, his partner, his first recruit. The red-haired boy grew taller but not softer, his sardonic humor sharpening into a defense against a world that had hurt him early and often. His mother, Lilly Ann, came to trust the Panning household enough to let Charlie spend most of his time there, and sometimes Wendy would catch her at the door, her eyes full of a gratitude that went beyond words.

"You saved him," Lilly Ann said once. "Not from his father—I finally did that myself—but from becoming bitter."

"Peter did that," Wendy said. "They're brothers now."

"That's more than anyone else ever did for us." Lilly Ann's voice was fierce. "Your son is going to be something extraordinary, Mrs. Panning. You mark my words. He's going to save people. He's already started."

***

On Peter's twelfth birthday, Barrie came with a gift. It was a small leatherbound journal, the pages blank and creamy, the cover embossed with a simple design of stars and wings.

"For your names," Barrie said. "The people you meet. The ones you help."

Peter turned the journal over in his hands. "What do I write in it?"

"Whatever you need to remember. Whoever you don't want to forget." Barrie's voice was soft. "Someday, Peter, you're going to encounter people who need to be remembered. People whose names might otherwise be lost. This is where you'll keep them. This is how they'll survive."

Peter opened to the first page and ran his finger down the blank paper.

"Why me?" he asked. "Why do you think I'll be the one to remember?"

"Because you came back." Barrie met his eyes steadily. "You found your way home when other children couldn't. That means you understand something most people never learn—that the lost can be found. That names matter. That windows should never be closed."

"And that dying *isn't* an adventure?"

Barrie was quiet for a moment.

"No," he said finally. "Dying *isn't* an adventure. Dying is the end of adventures. But living—really living, with your eyes open and your heart ready—that's the biggest adventure of all. And keeping others alive, keeping their stories alive..." He touched the journal. "That's a kind of immortality."

Peter closed the journal. Held it against his chest.

"Thank you," he said. "I'll use it. I'll use it to keep them all alive."

He was twelve years old. He had no idea what was coming.

But when the war arrived—when the lost boys stopped being metaphor and became flesh and blood and bone, dying in the mud of France—he would be ready.

The journal would be ready.

And the names would begin.

CHAPTER SIX

THE LOST BOYS

London, 1912

The year the boys turned fifteen, the boys discovered the abandoned greenhouse behind St. Andrew's.

It had been a grand thing, once. A Victorian glass palace where some wealthy family had grown orchids and oranges and other impossible things. Now the glass was mostly broken, the plants long dead, the iron frame rusted but still standing. The door hung crooked on its hinges, and ivy had crept through the cracks, reclaiming the space for wildness.

To the boys, it was perfect.

"This is it," Peter said, stepping through the door into the green-shadowed interior. "This is ours."

Charlie followed him in, hands in his pockets, eyes scanning the space with the wariness he never quite lost. "Ours for what?"

"For *us*. A place that's just ours. Where we can... I don't know, plan things. Be together."

"Be together doing what?"

Peter didn't answer immediately. He walked the perimeter of the greenhouse, touching the rusted frames, the broken glass, the vines that twisted through the ceiling like veins. Light filtered down in strange patterns, dappled and shifting, almost like—

"Like the prism," Charlie said, watching the same light. "In your nursery."

Peter smiled. "Yes. Exactly like that."

***

The greenhouse became their kingdom.

Peter and Charlie were the first, but others followed. Boys from school, from the neighborhood, from the hospital where Wendy still volunteered. Boys who were looking for something—a place, a purpose, a family that chose them instead of just inheriting them.

Billy Hawkins was the next to arrive.

He was fifteen too, cheerful to the point of absurdity, with a gap between his front teeth and a laugh that sounded like a rusty gate swinging open. He had been at Peter's school for years, but it was only now, in the greenhouse, that Peter really saw him—saw the way his optimism was armor, the way his jokes were shields, the way he believed in the future so fiercely because believing in the present was too painful.

"What is this place?" Billy asked, turning in a slow circle, taking in the broken glass and climbing vines.

"Headquarters," Peter said.

"Headquarters for what?"

"The Lost Boys."

Billy considered this. "Like the play?"

"Like the play. But real."

"And what do the Lost Boys do?"

"Look after each other." Peter gestured at Charlie, who was perched on an overturned crate in the corner, watching the exchange with sardonic amusement. "We don't leave anyone behind. We don't forget anyone. And when someone's lost, we help them find their way home."

Billy's grin widened. "That's very noble. Do we get biscuits?"

"Billy brought some," Charlie said. "He always has biscuits."

"Then I'm the perfect recruit." Billy flopped down onto a pile of old burlap sacks and pulled a tin from his pocket. "Ginger snaps. My mum makes them. She says if I'm going to wander off, I should at least not starve."

Peter took a biscuit and sat down across from him. "Tell me about yourself."

"What's to tell? I'm Billy. I like biscuits and football and a girl named Rose who laughs too loud and doesn't know I exist."

"She knows you exist," Charlie said. "She just pretends she doesn't because you stare at her like a lovesick puppy."

Billy threw a biscuit at him. Charlie caught it and ate it.

"Tell me about Rose," Peter said.

And Billy did. The words poured out of him like he'd been waiting for someone to ask—Rose's dark hair, Rose's laugh, the way she tilted her head when she was thinking, the way she smelled like lavender soap and summer afternoons. It was love in its purest adolescent form: hopeful, desperate, utterly convinced of its own eternity.

"We're going to be married," Billy announced. "After."

"After what?" Charlie asked.

"After... I don't know. School. Growing up. Whatever comes next." Billy shrugged, his gap-toothed grin stretching wide. "Doesn't matter what, really. As long as there's an 'after,' and Rose is in it."

Peter wrote it down in his journal—not the leather one from Barrie, which was still mostly empty, but a small notebook he'd started carrying everywhere.

Billy Hawkins. Fifteen. Optimistic. In love with Rose. Believes in 'after.'

It was the first entry of many.

***

They arrived together. The other boys might have assumed they were twins—same height, same way of standing, same way of speaking in tandem—but they were cousins. From Oklahoma. From the Choctaw Nation. They introduced themselves as Koi and Tiich.

"But our papers say Thomson and Thompson, the name the colonizers gave us at school," Koi—or rather Thomson—said.

"We're cousins, from the same tribe," Thompson said.

"But we look like twins," Thomson interject.

"I have the scar across my cheek when you need to tell us apart," Thompson said.

"We came over with the Wild West show two years ago."

"The show went back. We didn't."

"We heard there was a club. For lost boys."

"Are we lost?" Thompson looked at his cousin.

"Probably." Thomson looked back. "Everyone's lost, aren't they?"

"Not everyone."

"Everyone who matters."

Peter watched this exchange with fascination. "The Twins" spoke like two halves of the same thought, their sentences fitting together like puzzle pieces. When one paused, the other picked up. When one moved, the other mirrored.

"Do you do that on purpose?" Charlie asked. "The talking thing?"

"Do what on purpose?" Thompson asked.

"We're just talking," Thomson said.

"Like normal people," Thompson added.

"You're not normal people," Billy said cheerfully. "You're doing it right now. Watch—one of you start a sentence and stop in the middle."

"I don't—" Thompson began.

"—know what you mean," Thomson finished.

"See?"

The Twins looked at each other. And then, very deliberately, Thompson leaned forward and touched his forehead to Thomson's. Just for a moment. A gentle press, forehead to forehead, and then they separated.

"What was that?" Peter asked.

"Checking," Thompson said.

"That we're both real," Thomson explained.

"That neither of us is dreaming the other."

"We've done it since we were small."

"Grandpa says it's how we've always stayed connected."

"So we never get separated."

"So we always know the other one's there."

Peter felt something shift in his chest. He thought of shadows, of the part of you that goes to dark places so you don't have to. Of the desperate need to know that you were whole, that nothing essential had been lost.

"Still here," he said quietly.

The Twins looked at him.

"That's what you're saying, isn't it? With the forehead thing. 'Still here.'"

They considered the question. Then, in unison:

"Yes."

"And the answer?"

"Still together," Thomson said.

"Always," Thompson added.

Peter wrote it down: *Koi and Tiich. Thomson and Thompson. Cousins. The Twins. Ritual: forehead touch. "Still here." "Still together." "Always."*

It would become the Lost Boys' greeting. Their password. Their prayer.

They didn't know then how much they would need it.

***

Michael Davies came next. He was quieter than the others—a watcher, a recorder, a boy who seemed to exist slightly outside the moment, observing it for later use. He carried a notebook everywhere, battered and ink-stained and held together with string. He filled it with observations: the color of the light at different times of day, the exact pitch of Billy's laugh, the way Charlie's accent changed when he was nervous.

"What are you going to do with all that?" Peter asked once, peering over Michael's shoulder at a page covered in cramped handwriting.

"Write a book." Michael said it simply, like it was obvious. "Someday. A real book, with all of you in it. So people will know."

"Know what?"

"That we were here. That we mattered. That..." He gestured vaguely. "That there were boys like us, and we were friends, and we had adventures. Even if nobody else remembers, the book will."

"That's a lot of pressure for a notebook."

"It's not the notebook. It's what's inside it." Michael touched the cover. "Names. Stories. The things people say when they don't think anyone's listening. That's what lasts. Not the big stuff, the small stuff. The real stuff."

Peter thought about his own journal—the one from Barrie, still waiting for names he hadn't learned yet. "What's my small stuff?"

Michael flipped through pages, squinting at his own handwriting. "You watch windows. Even when there isn't one, you look for them. And you remember names—everyone's names, even people you've only met once. And when someone's hurt, you move toward them instead of away." He looked up. "You're a runner. You run toward things instead of away from them."

"Is that good?"

"I don't know yet." Michael closed the notebook. "I'll let you know when I write the ending."

Michael Davies. Writer. Observer. Wants to write a book about the Lost Boys.

Peter didn't write down what Michael had said about running toward things. He didn't need to. He already knew it was true.

***

Curly came last. No one knew his real name. When asked, he just shrugged and said "Curly" like it was obvious—which it wasn't, because his hair was perfectly straight. But the nickname stuck, and eventually, so did he.

He was bigger than the others, broad-shouldered and quiet, with hands that looked like they could break things but only ever seemed to fix them. He didn't talk much, but he was always *there*. Standing watch by the door, scanning for threats, positioning himself between the other boys and anything that might hurt them.

"He's like a sheepdog. Keeping the flock together," Billy said once.

"More like a guard dog," Charlie corrected. "The kind that doesn't bark until it's too late to run."

Curly heard them but didn't react. He just kept watching the door.

Later, Peter found out why. Curly had a sister—Emma, five years younger, the only family he really acknowledged. Their father had left years ago. Their mother worked double shifts at a factory. Curly had been the one to raise Emma, to protect her, to stand between her and a world that didn't care about little girls in poor neighborhoods.

"She writes me letters," Curly said one evening. It was the most he'd said in weeks. "Every day. Even when I see her at home, she still writes letters. Says it makes her feel like I'm not going anywhere."

"Do you write back?"

"Every time." His voice was soft. "She needs to know someone's listening."

Curly. Real name unknown. Protector. Has a sister named Emma who writes letters.

Peter was starting to understand what Michael meant about the small stuff. The details that made a person real. The fragments that added up to a life.

***

Geoffrey appeared at the greenhouse door in the spring of 1914. He was small and scrawny and practically vibrating with determination.

"I want to join," he said.

The Lost Boys looked at each other.

"Join what?" Billy asked.

"The group. The Lost Boys. I know what you do—you look out for each other. You don't leave anyone behind." Geoffrey's chin jutted out. "I want to be one."

"You're a bit young, mate," Charlie said, not unkindly.

"I'm almost fifteen."

"You're thirteen and a half. I asked your sister."

The boy's face reddened. "Age doesn't matter. What matters is... is..." He struggled for words. "Is whether you'll stay. Whether you'll be there when it counts."

Something in his voice made Peter look closer. There was a bruise on the boy's arm, half-hidden by his sleeve. A familiar shape. A familiar story.

"What's your name?" Peter asked.

"Geoffrey." A hesitation. "Just Geoffrey."

"All right, Just Geoffrey. You can stay."

"Really?"

"Really. But you have to follow the rules."

"What rules?"

"Don't leave anyone behind. Write down the names. And when someone says 'still here,' you say..."

"'Still together,'" Geoffrey finished, his eyes bright. "I know. I've been listening."

Charlie caught Peter's eye. His expression said: *he's too young for this.*

Peter's expression said: *so were we.*

Geoffrey. Youngest. Desperate to belong. Following rules he doesn't fully understand.

And that was how Geoffrey became the youngest Lost Boy—the one who would grow up too fast, learn too much, and keep the ledger when Peter couldn't.

The one who would die at the Somme, three weeks after his birthday, still too young for everything except the grave.

But that was later. That was after.

For now, he was just a boy in a greenhouse, surrounded by other boys, learning what it meant to belong.

***

The summer of 1914 came in hot and golden.

The Lost Boys spent it in the greenhouse, playing cards and telling stories and arguing about things that didn't matter. Billy talked about Rose constantly. The Twins developed an elaborate system of signals that no one else could understand along with their Choctaw language. They would sometimes hold the dirt in their hands; feel it, smell it, and lay their ears to the ground. Michael filled notebook after notebook. Curly stood watch. Geoffrey tried too hard and talked too much and slowly, slowly, started to relax.

And Charlie—"Slightly," they'd started calling him, because he was always slightly miserable, slightly sarcastic, slightly certain he was smarter than everyone else—sat in his corner and watched and made cutting remarks that were really, if you listened closely, a kind of love.

"This is good," Peter said one evening, as the light faded and the greenhouse filled with shadows. "This is what I wanted."

"What you wanted for what?" Charlie asked.

"For the lost ones. For the boys who don't have anywhere else to go." Peter looked around at his friends—his brothers, though none of them shared blood. "I wanted to build something. A family."

"Very philosophical of you."

"I mean it. When I was—" He stopped. Started again. "When I was away. Before. I saw what happens to children who don't have anyone looking for them. They forget how to hope. They forget how to go home. I never want any of us to forget that."

The greenhouse was quiet. Even Billy's relentless cheer didn't break the silence.

And Charlie—Slightly—looked at Peter with something in his eyes that might have been understanding, or might have been love, or might have been the knowledge that this couldn't last forever.

"Nothing will pull us apart."

"Something always does." Charlie's voice was gentle. "That's what the world does. It takes things. It takes people. And all we can do is hold on as long as we can and remember each other after."

"Then we'll remember. We'll write it all down. We'll never let anyone be forgotten."

"Is that a promise?"

"It's the only promise that matters."

***

The news came in August.

War. Germany. France. The whole world catching fire, and Britain standing at the edge trying to decide whether to leap into the flames.

The boys gathered in the greenhouse, passing a newspaper between them, reading the headlines in disbelief.

"They're saying it'll be over by Christmas," Billy said. His voice was strange—not cheerful, but trying to find its way back there. "A grand adventure. Show the kaiser what's what."

"It won't be over by Christmas," Charlie said flatly. "Wars never are."

"How would you know?"

"History. Books. The basic inability of politicians to solve anything quickly." Charlie shook his head. "This is going to be bad. Really bad. And they're going to need soldiers."

The word hung in the air.

Soldiers.

Peter looked around at his Lost Boys. Billy, with his optimism and his Rose and his faith in 'after.' The Twins, who had never been apart, whose foreheads still touched every morning like a prayer. Michael, scribbling in his notebook, trying to capture something that couldn't be captured. Curly, standing guard, watching the door. Geoffrey, too young, too eager, already burning to prove himself.

And Charlie. Slightly. The sardonic voice, the watchful eyes, the boy whose mother had named him and loved him and stood in doorways trying to stop him from going.

They were going to enlist. Peter could see it in their faces. All of them, even the ones who were too young, who would lie about their ages and slip through the cracks in a system that needed bodies more than it needed truth.

They were going to war.

And Peter was going with them.

"Still here," he said.

"Still together," they answered.

And the world tilted toward darkness.

Chapter Seven

The Journal

London, Autumn 1914

The recruitment posters were everywhere. Kitchener's face, stern and mustachioed, pointing directly at you from every wall, every window, every surface that could hold paper and paste. YOUR COUNTRY NEEDS YOU. The words followed Peter through the streets, into his dreams, across the breakfast table where his mother and father sat in careful silence.

JOIN YOUR CHUMS AT THE FRONT. DON'T BE LEFT BEHIND. WOMEN OF BRITAIN SAY GO.

"They're making it sound like a holiday," Charlie said, scowling at a poster as they walked home from the greenhouse. "A grand adventure. Jolly good fun for all the lads."

"Isn't it?" Billy was walking backward ahead of them, his face bright with something that wasn't quite excitement but wanted to be. "I mean, it's our chance, isn't it? To do something. To matter."

"We already matter."

"Not like this. Not to the whole country. Not to *history*." Billy spread his arms wide, nearly colliding with a lamppost. "Our grandchildren will ask what we did in the war, and we'll have stories. Real stories. Not just..." He gestured vaguely. "Sitting in a greenhouse talking about feelings."

"I like talking about feelings," Michael said quietly. He was scribbling in his notebook even as he walked, some observation about the posters or the crowds or the strange fever that had gripped the city. "Feelings are what matter. When the war's over, no one will remember the battles. They'll remember how people felt."

"That's rubbish. They'll remember who won."

"Will they? A hundred years from now?" Michael shook his head. "They'll remember the letters. The poems. The small moments between the big ones. That's always what lasts."

Peter listened to them argue but said nothing. He was watching the men in uniform—soldiers, or soldiers-to-be, moving through the crowds with a purpose that hadn't existed six months earlier. Some of them looked proud. Some looked terrified. Some looked like they were already somewhere else, their bodies going through the motions while their minds traveled ahead to places they couldn't imagine.

He recognized that look.

He'd worn it himself, once. Coming back from the "island." Returning to a world that felt too solid, too bright, too *here* after so long in the dark.

They're going to Neverland, he thought. *Only this Neverland has guns.*

***

The decision happened in pieces.

First Billy, who couldn't wait, who lied about his age the moment he turned seventeen and presented himself at the recruitment office with his gap-toothed grin and his unshakeable faith in 'after.'

"Rose knows," he told them that night, practically glowing. "She cried, but then she kissed me—actually *kissed* me—and said I was the bravest boy she knew. We're going to be married when I come home. She promised."

"You're an idiot," Charlie said.

"Probably. But I'm an idiot in uniform now." Billy pulled out a photograph of Rose laughing at something beyond the frame, her dark hair catching the light. "I'm going to keep this next to my heart. Every battle. Every march. She'll be right there with me."

He tucked the photograph into his breast pocket. Patted it twice.

It was the first time Peter saw his ritual. It wouldn't be the last.

***

The Twins went together, of course. They appeared at the greenhouse in matching uniforms, looking for all the world like a single person reflected in a mirror. Thompson had a bruise on his cheek from the recruitment sergeant who'd tried to separate them.

"We told him we were a package deal," Thomson explained.

"That we're fighting for a country that took our grandparents' land," Thompson said.

"We aren't even recognized as a citizen in our own country," Thomson added.

"So, it's both of us or neither," Thompson added.

"He didn't like that."

"But we're very persuasive."

They touched foreheads, and Peter felt something crack in his chest. They were eighteen. They thought they were invincible. They thought the war would be a grand adventure and they'd come home with stories.

They had no idea.

None of them did.

***

Michael enlisted quietly, without fanfare.

"Someone has to write it down," he said when Peter found him at the recruitment office. His notebook was already open, already being filled it with observations. "The truth. Not the propaganda, not the glory—the real thing. What it actually looks like when boys go to war."

"You could write it from here. From London."

"No, I couldn't." Michael met his eyes steadily. "You can't write about something you haven't seen. I learned that from you. You never try to explain the 'island.' You just carry it with you. That's the only honest way to hold a story."

"I don't want to hold the *island*. I want to forget it."

"But you can't. And that's what makes you true." Michael closed his notebook. "I'm going to be true. Whatever that costs."

Michael Davies. Writer. Going to war to find the real story.

Peter wrote it down in his journal—the leather one from Barrie, filling up faster now with names he knew he might lose.

***

Curly didn't say anything to anyone. He just appeared one day in uniform, his broad shoulders filling out the khaki, his face as unreadable as ever. When Peter asked why, he pulled out a letter. It was creased and tear-stained, clearly read a hundred times.

"Emma," he said. "She wrote that she was proud of me. Before I even enlisted. She said she knew I'd go because I always take care of people, and now I get to take care of the whole country."

"She's twelve years old," said Peter.

"She's the smartest person I know." Curly tucked the letter back into his pocket. "She said she'd write me every day. And I said I'd write back every day. As long as there are letters, she'll know I'm *still here*."

***

Geoffrey tried three times.

The first time, they laughed at him. Sent him home with a pat on the head and a reminder that the war needed men, not children.

The second time, they were less kind. An officer called him a "desperate little fool" and threatened to write to his mother.

The third time, he arrived with papers he'd clearly forged, claiming to be eighteen, and somehow—somehow—it worked. The recruitment officer was tired, the line was long, the war was hungry for bodies. Geoffrey slipped through the cracks.

"They took me," he said, breathless, appearing at the greenhouse that evening. "They finally took me. I'm going. I'm really going."

"You're a baby," Peter said.

"Fourteen in November."

"You're *thirteen.*"

"And you're seventeen, which is also too young, so don't lecture me about rules." Geoffrey's chin jutted out. "I'm a Lost Boy. We don't leave anyone behind. And I'm not being left behind."

There was nothing Peter could say to that. Nothing that wouldn't be hypocrisy.

Geoffrey. Sixteen. Too young. Going anyway.

***

And then there was Charlie.

He waited longer than the others. Peter found him one evening at his mother's flat—not the cramped, angry place from childhood, but a small clean room in a better neighborhood. Lilly Ann had finally left Charlie's father, had clawed her way to something like safety. She was working as a seamstress now, her small, fierce hands creating beauty out of fabric and thread.

When Peter arrived, Charlie was sitting at the kitchen table, and Lilly Ann was crying.

"Three hours," Charlie said. His voice was strange. Hollow. "She's been crying for three hours. I keep telling her I'll be fine, but she won't—"

"I'll cry as long as I bloody well want to," Lilly Ann said, wiping her face with her apron. She was tiny—barely five feet tall, red-haired like her son, with the same sharp eyes. "My boy is going to war, and I'm allowed to cry about it."

"Mum—"

"I stood in the doorway," she said, turning to Peter. Her voice was fierce, almost angry. "When he told me he'd enlisted. I stood in the doorway and I wouldn't move. I thought... I thought if I just stood there, he couldn't leave. He'd have to stay."

"She weighs about eight stone," Charlie said quietly. "I had to pick her up and move her."

"And he did. My own son. Picked me up like I was nothing and set me aside." Lilly Ann laughed—a broken sound, half sob. "That's when I knew. He's not my baby anymore. He's a man. And men go to war, and mothers stand in doorways crying, and nothing ever changes."

Peter looked at this small, fierce woman. Lilly Ann Morton had survived a brutal husband, had raised a son alone, had fought for every inch of the life she'd built. He understood, suddenly, where Charlie got his sharp edges. Where he got his strength.

"He'll come back," Peter said. "I'll bring him back myself."

Lilly Ann stared at him. Her eyes were red-rimmed, her face blotched with tears, but her gaze was steady.

"You promise?"

"I promise."

"Then you'd better keep that promise, Peter Panning." She pointed a finger at him, the same way she might have pointed at her son. "You bring my Charlie home to me. Because if he dies over there, I swear to God, I'll find a way to bring him back just so I can kill him myself."

"Mum," Charlie said weakly.

"Don't 'mum' me. I mean it. Both of you. You come home." She grabbed the two of them, one in each arm, squeezing with surprising strength. "You come home, or I'll never forgive you."

Peter hugged her back. She smelled like tea and soap and the particular desperation of mothers saying goodbye. Lilly Ann pulled back and looked at him with something like recognition.

Lilly Ann Morton. Charlie's mother. Small. Fierce. Stood in the doorway and tried to stop the war.

She couldn't.

No one can.

***

The night before they left, Billy found Rose.

He came back to the greenhouse just as the sun was setting, his face transformed into something Peter barely recognized. There was joy, yes—but also fear, and hope, and the particular gravity of someone whose life had just become real.

"She said yes."

The boys looked at him.

"Rose. I asked her... I told her we might not... that I didn't know when I'd be back, or if... but I asked her anyway. And she said *yes.*"

"Asked her what?" Geoffrey demanded.

"To marry me. When I come home. For real, this time. Not just—not just 'after,' but *really.*" Billy's hands were shaking. He pulled out the photograph, handling it like a holy relic. "She gave me this. Told me to keep it next to my heart. She said... she said every time I look at it, she'll know. Wherever she is, she'll feel me looking, and she'll laugh."

"She laughs too loud," Charlie said. But his voice was soft.

"I know. That's why I love her." Billy tucked the photograph into his breast pocket. Patted it twice. "She's going to be my wife, lads. When this is over. She's going to be Mrs. Rose Hawkins, and we'll have a house with a garden, and children who laugh too loud like their mum, and—" His voice cracked.

Peter put a hand on his shoulder. "We'll get you home. We'll get you back to Rose," he said.

"Promise?"

And for a moment, his gap-toothed grin was back—the same one he'd worn the day he'd stumbled into the greenhouse, all optimism and biscuits and unshakeable faith in 'after.'

***

The platform was chaos. Soldiers everywhere, families clinging, brass bands playing songs meant to sound cheerful but landing like dirges. The Lost Boys found each other in the crowd—they always found each other—and stood in a tight circle, their uniforms too new, their faces too young, their hands searching for something to hold.

"This is it, then," Billy said. He was grinning, but his eyes were wet. "The grand adventure."

"Don't be dramatic," Charlie muttered. "It's a train ride. We're not dying yet."

"Yet," The Twins said in unison.

"That's not helpful."

Curly was scanning the crowd, looking for threats even here. Geoffrey was vibrating with nervous energy, his too-big uniform making him look even younger than he was. Michael had his notebook out, scribbling furiously.

"What are you writing?" Peter asked.

"Everything. The sound of the band. The smell of coal and perfume. The way that mother over there is holding her son like she'll never let go, even though she already has." Michael's voice was steady, but his hand shook. "Someone has to remember this."

"We all will."

"No. Memory fades. Words don't." He looked up at Peter. "That's why your book matters—the one Mr. Barrie gave you. You have to write everything down. Every name. Every face. Every small thing that makes a person real."

"I will."

The whistle blew. Final boarding.

Peter looked at his Lost Boys. His brothers. The boys who had sat with him in a broken greenhouse, planning adventures that had seemed impossible and now were all too real.

"Still here," he said.

"Still together," they answered. All of them. In unison.

He saw his mother at the edge of the platform, her face pale and still, her hands clasped in front of her like a prayer.

He remembered the night he'd told them. The nursery. The prism catching the last of the evening light. His mother at the window, his father standing in the doorway, both of them already knowing. He could see it in their faces. Parents always knew.

"I have to go with them," Peter said. "They're my boys. I can't let them go alone."

Wendy didn't speak. Her hands were very still in her lap.

Willem cleared his throat. "You're not old enough. Not technically. The recruitment office—"

"Will take me. They're taking everyone." Peter looked at his mother. "You know I have to. You've always known."

The silence stretched. The prism turned slowly, scattering colors across the walls—the same walls where Peter had chased rainbows as a child, before the island, before any of it.

"I kept the window open for three years," Wendy said finally. Her voice was steady, but something beneath it was breaking. "I kept it open because I believed you'd come home. I never stopped believing."

"I know."

"If you go—" She stopped. Started again. "If you go, I'll keep it open again. However long it takes. But Peter—" Her eyes held everything she couldn't say. "You have to come back. You have to try. Promise me you'll try."

"I promise."

Willem crossed the room and put his hand on Peter's shoulder. His grip was firm, the way it had been when Peter was small and learning to walk again after coming home from the dark.

"You're going to look after them," Willem said. It wasn't a question. "The lost ones. That's why you're going."

"Someone has to."

His father's voice was rough. "Then go. Go, and write to us, and keep your head down, and come home. In that order."

Peter hugged them both—his mother who had never stopped watching, his father who had never stopped believing her. He breathed in the smell of the nursery, the lavender and dust and particular warmth of the only home he'd ever wanted to return to.

***

He raised his hand—not a wave, but a salute—and something passed between them that words couldn't convey.

I'll come back. I'll keep the window open from my end.

She nodded. She understood.

And then he was on the train, pressed against the window, watching London fall away behind him—the rooftops and chimneys, the distant parks, the window at 14 Wimpole Street, where a prism caught the light and scattered rainbows that no one was there to see.

He opened the leather journal and found the first blank page.

Day One.

Left London this morning. Mama and papa at the station. She didn't cry. I think that might have been harder than if she did.

Michael is in my car, and Charlie too. Billy, The Twins, Curly, Geoffrey—we're all together. We're trying to pretend we're not terrified. I don't think any of us are very good at it.

I keep thinking about what Barrie said once—that to die would be an awfully big adventure. I used to think that was brave. Now I'm not so sure.

I think maybe the adventure isn't dying.

I think maybe the adventure is staying alive.

Window open. Heading toward the dark.

Names to collect: Michael Davies. Charlie Morton. Billy Hawkins. Thomson. Thompson. Curly. Geoffrey. Peter Panning.

Lost boys, all of us.

But I'm going to bring us home.

He closed the journal. Pressed it against his chest.

Outside the window, the countryside was slipping past, green and gold and impossibly peaceful. Somewhere ahead, the guns were waiting. The mud was waiting. The crocodile was waiting, tick-tick-ticking away the seconds until it could swallow them whole.

But for now, they were together.

For now, that was enough.

Interlude I

The First Lost Boys

London, Autumn 1915

The house was too quiet. Wendy had never noticed how much space Peter took up—not just physically, but in every other way. The sound of his footsteps on the stairs. The particular rhythm of his breathing in the next room. The way he would appear in doorways without warning, watching, always watching, as if making sure the world was still where he'd left it.

Now the doorways stayed empty. The stairs stayed silent. And Wendy moved through the house like a ghost, touching the places where Peter used to be.

His room was exactly as he'd left it. She hadn't changed anything—couldn't bring herself to, even though Willem had gently suggested that dusting the shelves might help. The bed was made (he'd made it himself that last morning, corners tucked with military precision), and the window was open (always open, always), and the prism still hung from its chain, still scattering rainbows when the light was right.

She sat there every evening in the rocking chair. Watching the light change. Waiting.

Come home, she thought. *Please come home.*

The letters arrived sporadically. They were censored, carefully cheerful, saying nothing and everything at once. Peter wrote about the weather. The food. The "interesting characters" in his unit. He never mentioned mud or shells or the names of boys who weren't there anymore.

But between the lines, Wendy read the truth.

Her son was in hell. And all she could do was keep the window open.

***

The first soldier arrived in June. He wasn't what Wendy expected. Soldiers, in her imagination, were strong and proud and whole—the boys on the recruitment posters, striding toward golden horizons with rifles on their shoulders.

This man was none of those things.

He was sitting on her front steps when she returned from the hospital, huddled in an ill-fitting uniform that had seen too much wear. His hands were shaking. His eyes were fixed on something she couldn't see—some horizon that existed only in memory, some moment he couldn't escape.

"Can I help you?" Wendy asked gently.

He flinched. Actually flinched, as if her voice were a shell exploding. Then, slowly, his eyes focused on her face.

"The window," he said. His voice was rusty, unused. "Someone said... there was a window. A house where the window's always open."

Wendy's heart stopped.

"Who told you that?"

"A boy. Said if I ever got lost, I should look for the house with the open window. Said the lady there would help."

"What was his name? The boy who told you?"

"Panning." The soldier's face struggled to remember. "Peter Panning. He was... he was kind. Came to visit the wounded. Wrote down names. Said no one should be forgotten."

Wendy sat down on the steps beside him. She didn't touch him—she'd learned that much at the hospital, learned that some wounds made touch unbearable—but she sat close enough that he could feel her presence.

"That's my son," she said. "Peter."

The soldier looked at her. Really looked for the first time. And something in his face cracked—some wall he'd been holding up, some defense that suddenly wasn't necessary.

"He said you'd keep the light on. Even in the dark."

"I do. I always do."

"He said—" The soldier's voice broke. "He said you'd help me remember who I am. Because I don't—I can't—I don't know anymore. I don't know who I was before. All I can see is..."

He stopped. Couldn't finish.

Wendy reached out, very slowly, and touched his hand. He didn't flinch this time. He just looked at her fingers on his, as if trying to remember what human contact felt like.

"What's your name?" she asked.

"Samuel. Samuel—" He struggled. "I can't remember my last name. Isn't that strange? I can remember the sound of shells. I can remember the smell of..." He shuddered. "But I can't remember my own name."

"Samuel is enough. For now." Wendy stood, gently pulling him to his feet. "Come inside. I'll make tea. And you can stay as long as you need."

"I don't... I can't pay—"

She met his eyes steadily. "I'm not asking for payment. I'm asking you to come inside. My son sent you to me because he knew I'd help. He's over there, fighting, and I'm here, and I can't do anything for him. But I can do something for you. Will you let me?"

Samuel stared at her for a long moment. Then, slowly, he nodded.

"Still here," he whispered. It sounded like a question.

Wendy understood.

"Come inside, Samuel."

***

More came after that.

Not many at first—a trickle, then a stream. Word spread through the hospitals, the convalescent homes, the places where broken soldiers were sent to either recover or be forgotten. There was a house on Wimpole Street. A woman who didn't ask questions. A window that never closed.

Wendy watched them arrive. Thomas, who couldn't speak and flinched at sudden noises. Edward, whose hands shook so badly he couldn't feed himself. James, who wept every night for a brother who hadn't come home. Michael (not *Peter's* Michael, she told herself, not yet), who had been at the Somme and couldn't remember anything before or after, only the endless noise and the endless mud.

They were so young. That was the thing that broke her heart every time. They had boys' faces, but their eyes were ancient. They had survived something that no training, no poster, no grand speech could have prepared them for. And now they were here, in her house, trying to remember how to be human.

"You're running a hospital," Willem said one evening. He wasn't criticizing—Willem never criticized her choices—but there was worry in his voice.

"I'm running a home." Wendy was folding bandages. Her hands never stopped moving anymore. "They don't need more hospitals. They've had enough of hospitals. They need somewhere to *be*."

"You can't save them all."

"I know."

"Some of them won't get better."

"I know that too."

Willem sat down beside her. His hair was fully gray now, his face lined with the particular exhaustion of a man who had spent a year watching his wife pour herself into something that might never be finished.

"You're thinking about Peter," he said.

"I'm always thinking about Peter."

"He'll come home."

Wendy's hands stilled. "Will he? Even if he does—even if his body comes home—will *he*? Will the boy we raised, the boy who watches windows and writes names and can't bear to leave anyone behind?" She looked at the bandages in her lap. "Or will he come home like Samuel? Like Thomas? Lost inside himself, looking for something that was taken from him?"

Willem had no answer. There wasn't one.

"All I can do is be ready," Wendy finally said. "Keep the window open. Keep the light on. And when he comes home—whoever he is, whatever he's become—I'll be here."

"You're doing what Peter does," Willem said quietly. "Finding the lost ones. Giving them a place to land."

"I learned it from him. Or he learned it from me. I'm not sure anymore." Wendy smiled tiredly. "We're fighting the same war, Peter and I. Just from different sides of the Channel."

***

The letter from Peter arrived in August. It was different from the others—longer, more careful, as if he'd spent a long time choosing his words.

Dear Mama,

I've been thinking about the boys here. The ones I write for, the ones I sit with. The ones who are going home and the ones who aren't.

Some of them are so lost. Not just hurt—lost. Like they've forgotten where they came from, who they are, what they're supposed to want. The doctors call it shell shock. I call it something else.

I call it forgetting how to find the window.

I've been telling them about you. About the house on Wimpole Street. About how you kept the window open for me, for years, and never gave up. I've been telling them that if they can't find their own window, they should find yours. That you'll help them remember.

I hope that's all right. I hope I haven't sent too many. I just need to know that someone's there for them. That when they get back to England, there's a light waiting. That they're not just being sent home to disappear.

The war is—

I can't write about the war. The censors would cut it anyway. But I want you to know: I'm still here. I'm still writing the names. And every time I do, I think of you, and the prism, and the way the light scatters but doesn't disappear.

Keep the light on, Mama. I'll find my way home.

Your son,

Peter

Wendy read the letter three times. Then she folded it carefully, pressed it to her chest, and went to the nursery.

The window was open. The prism was catching the afternoon light. And somewhere, across the water, in a trench that smelled of death and desperate hope, her son was keeping his own light burning.

"I'll keep it open. I'll keep it lit," she whispered. "For you. For all of them."

The shadows shifted. The light scattered.

And in a house that had once been too quiet, filled now with the murmurs of broken boys learning to be whole, Wendy Panning kept watch.

Always.

PART II

NO MAN'S LAND/NEVERLAND

Interlude II

The Second Start

Second star to the right, and straight on till morning.

That's what the story says. That's how you find Neverland—the island where children never grow old, where the trees whisper secrets, where lost boys build kingdoms out of imagination and defiance.

But the story doesn't tell you what happens when morning never comes.

It doesn't tell you about the other island that waits across the water, past the ships and the shells and the drowned hopes of a thousand boys. The one that has no name except the name the soldiers gave it:

No Man's Land.

***

Once, there was a boy who could fly.

He flew because he believed he could—because he had forgotten that gravity was supposed to hold him, that the ground was supposed to claim him, that boys who leapt from windows were supposed to fall.

He flew to an island where the rules were different. Where children made the laws. Where the only adults were villains, and villains could always be defeated.

He flew away from the dark thing that had taken him, the place without light, the years of forgetting. He flew toward a mother who had kept the window open.

He flew home.

But some boys never stop flying.

Some boys learn that the air is safer than the ground—that movement is survival, that stillness is death. Some boys spend their whole lives looking for the next window, the next light, the next door that might lead somewhere better.

And some boys fly straight into thunder, because running toward danger is the only language they know.

***

In 1915, the boys who went to war were children themselves. Oh, they didn't think so. They thought they were men—eighteen, nineteen, twenty—with futures stretching before them like roads to be conquered. They had mustaches they'd cultivated for months. They had sweethearts who pressed flowers into their hands at railway stations. They had mothers who packed their bags with extra socks and pretended not to weep.

They had no idea what they were walking into.

How could they? The wars they knew were from stories—tales of cavalry charges and glorious last stands, of honor and sacrifice and enemies who played by rules. They had grown up on Kipling and Tennyson, on paintings of Waterloo, on the certainty that British boys were made of something finer than other nations' sons.

They believed—because they had been taught to believe—that war was an adventure.

To die would be an awfully big adventure.

A playwright had written that line for a boy who could fly, a boy who never grew up, a boy who faced pirates and crocodiles and the villain of his own imagination with nothing but courage and cleverness and the absolute conviction that he would win.

The boys of 1915 had seen that play. They had cheered when Peter Pan defeated Captain Hook. They had believed, in the secret chambers of their hearts, that they were Peter too—heroes waiting for their moment, boys who would fly toward danger and emerge triumphant.

They didn't know that Neverland had another name.

They didn't know that No Man's Land was waiting for them.

Chapter Eight

Into the Dark

France, April 1915

The train stopped at a place that had no name.

Or if it had a name, no one told them. The boys stumbled out onto a platform that was less a platform than a patch of churned earth, surrounded by buildings that had been beautiful once and were now just walls with memories of roofs. The sky was gray. Everything was gray—the ground, the buildings, the faces of the soldiers who watched them disembark with eyes that held nothing at all.

"Welcome to France, lads," a sergeant called out. He was older than Peter had expected. Thirty, maybe forty, with a face like crumpled paper and a voice that sounded like it had been dragged through gravel. "Form up. Single file. Keep your mouths shut and your eyes open."

The Lost Boys found each other in the crowd. They always found each other.

Billy was grinning—of course he was—but the grin was different now. Tighter, like he was holding it in place through sheer force of will. The Twins stood shoulder to shoulder, their hands brushing, not quite holding. Michael had his notebook out, scribbling something Peter couldn't see. Curly was scanning the platform, cataloguing threats, positioning himself between Geoffrey and the unknown.

And Charlie—Slightly—was watching Peter with those sharp, sardonic eyes.

"Second thoughts?" he asked quietly.

"No."

"Liar."

"Maybe." Peter looked at the gray sky, the gray buildings, the gray men waiting to lead them somewhere worse. "But second thoughts don't matter now, do they? We're here. There's no going back."

"There's always going back. Desertion. Cowardice. A convenient injury."

"Is that what you're planning?"

Charlie's jaw tightened. "I'm planning to survive. Whatever that takes."

"Then we'll survive together."

Something flickered in Charlie's eyes. It might have been gratitude, or might have been fear, or might have been the recognition that survival was no longer something either of them could promise.

"Still here," he said.

"Still together."

And they followed the sergeant into the gray.

***

They marched for three hours before they heard the guns.

At first, Peter thought it was thunder. The sound rolled across the flat, ruined landscape like a storm approaching—low and continuous, punctuated by sharper cracks that made the air itself seem to flinch. Basic training had not prepared them for this.

"That's the front," the sergeant said. He hadn't told them his name. Maybe names didn't matter here. "Five miles, give or take. You'll get used to it."

"How do you get used to that?" Geoffrey asked. His voice was too high. Too young.

"You don't. You just stop noticing." The sergeant's eyes swept over their clean uniforms, their unmuddied boots, their faces that still held the shape of boys. "Enjoy the quiet while you can. Tomorrow, you'll be close enough to feel it in your teeth."

They kept marching.

The landscape changed as they walked. Fields gave way to cratered earth. Trees gave way to stumps, splintered and blackened, reaching toward the sky like broken fingers. Houses gave way to rubble, and then to nothing at all—just emptiness where something used to be.

And everywhere, the smell.

Peter had been prepared for blood. For death. For the stench of wounds and decay that he'd learned to recognize at the hospital.

He hadn't been prepared for this.

It was a smell that didn't have a name—or rather, it had too many names, all of them true at once. Mud and rot. Cordite and copper. Latrine buckets and chloride of lime. Bodies that couldn't be retrieved and bodies that had been retrieved too late. The particular sweetness of gangrene and the particular sharpness of fear.

The smell of a world that had forgotten what clean meant.

"Breathe through your mouth," Charlie muttered. His face was pale. "It helps."

"Nothing helps." Michael had stopped writing. His notebook hung limply in his hand. "This is... I can't... there aren't words for this."

Peter's voice was harder than he intended. "Then find them. Find them, because someone has to remember. Someone has to write it down."

Michael looked at him. Something passed between them—an understanding, a commission, a burden shared.

"I'll try," Michael said.

"That's all any of us can do."

***

They reached the communication trenches at dusk.

The diagrams in training had been neat and clean—firing trench, support trench, reserve trench, all connected by tidy zigzagging lines. The drawings had been geometric, almost elegant, like something from an architecture textbook.

The reality was a wound in the earth.

Peter followed the soldier ahead of him through a gap in a ruined wall, then down into the dark, down into the mud, down into a smell so thick and complex it seemed to have weight. The trench walls rose on either side, shored up with rotting planks and sandbags that oozed brown water. The floor was a river of mud that sucked at his boots with every step, reluctant to let him go.

"Keep moving," the guide called back. "Don't stop. Don't look up. Don't touch the walls."

Peter didn't ask why. He could guess.

Behind him, he heard Billy stumble and curse. Heard one of The Twins catch his cousin's arm. Heard Geoffrey's breathing come fast and shallow, on the edge of panic.

The communication trench wound through the earth like an intestine, twisting back on itself in sharp angles designed to limit the damage from shells. Every few yards, the walls changed character—sometimes reinforced with corrugated iron, sometimes bare earth, sometimes nothing but sandbags piled haphazardly against collapse.

And everywhere, in every surface, there were things that shouldn't have been there.

A boot, protruding from the wall at knee height.

A hand—just a hand, fingers curled, gray-green with decay—reaching out from the mud as if waving hello.

A ribcage, exposed by erosion, the bones picked clean and gleaming dully in the half-light.

The trenches were built through graveyards.

Or perhaps the graveyards had grown up around the trenches. Peter couldn't tell anymore which had come first—the living or the dead.

Behind him, he heard Thompson's sharp intake of breath, half-gasp, half-sob. Peter turned.

The Twins had stopped. Thompson was staring at the hand in the wall, his face white, his whole body trembling. Thomson had grabbed his arm, was pulling him close, was pressing their foreheads together in that ritual they'd done since childhood.

They separated. Thompson's breathing was steadier now, but his eyes—his eyes were different. Older. As though he'd aged years in the span of seconds.

Peter understood.

They were all going to age like that. All going to look at things that would carve years into their faces, decades into their eyes. By the time this war was over—if it was ever over—none of them would be boys anymore.

He took out his journal. In the dim light, barely able to see the page, he wrote:

The trenches are built through the dead. The dead are part of the walls. We walk on them. We sleep beside them. We will become them.

The Twins held each other and said the words.

I don't know how long that will be true.

***

The firing trench was worse. It was narrower here, the walls pressing close, the sky reduced to a thin strip of gray above. Men lined the walls—some standing, some sitting, some curled into niches carved from the earth like animals in burrows. They looked up as Peter's group passed, their eyes tracking the new arrivals with dull curiosity.

No one spoke. No one smiled. No one offered the cheerful welcomes Peter had expected from the recruitment posters, the training sergeants, the letters home from boys who had gone before.

Those letters had been lies. Peter understood that now.

Not intentional lies, perhaps. Just the only truth that could survive the journey home. The censors would have cut anything real. And even without censors, how could you write it? How could you put into words the smell of a trench in April—the rotting sandbags, the latrine buckets, the bodies that couldn't be retrieved, the feet that had begun to decay inside boots that couldn't be removed?

How could you tell your mother that the adventure she had sent you to was a grave that hadn't finished filling?

"Here." The guide stopped at a section of trench that looked no different from any other. "This is your sector. Find a spot, settle in. Officer will come round at dusk to assign watches."

He was gone before Peter could ask any questions.

The Lost Boys stood in the narrow space, pressed together, looking at each other with faces that didn't know what expression to wear.

"Right," Billy said. His voice cracked on the word. "Not so bad, is it? We've got walls. We've got..." He looked around desperately. "Mud. Plenty of mud. And the company's good."

"Billy. Stop." Charlie's voice was gentle.

"I'm just saying—"

"I know what you're saying. And I love you for trying. But stop." Charlie put his hand on Billy's shoulder. "We're allowed to be scared. It doesn't make us cowards. It makes us sane."

Billy's gap-toothed grin flickered, steadied, and found its shape again. "Since when are you the comforting one?"

"Since about thirty seconds ago. Don't get used to it."

"I wouldn't dream of it."

A soldier passing by—old, hollow-faced, with eyes like stones that had seen too much water—heard them and laughed. Not a real laugh. A sound like something dying.

"New boys," he said. "I can always tell. You still think comfort matters."

"Doesn't it?" Peter asked.

The soldier looked at him. Looked through him, really, to somewhere Peter couldn't see.

"Ask me again in a month," he said. "If you're still here."

He walked on. His footsteps made no sound in the mud.

Peter wrote in his journal:

A soldier laughed at us. Said comfort doesn't matter. His eyes were empty.

I wonder if our eyes will look like that. I wonder how long it takes.

I wonder if there's a way to keep something alive inside, even when everything outside is dying.

***

The dugout was a hole in the ground.

That was the kindest way to describe it. A cavity scraped from the trench wall, shored with rotting timbers, large enough for eight men if they didn't mind sleeping in a pile. It smelled of damp earth and sweat and something older—something that had soaked into the ground and would never come out.

The Lost Boys claimed it for their own. They spread out their kits, found corners, arranged themselves in the particular geometry of boys who had learned to share space.

Billy took the spot nearest the entrance—he'd always liked being able to see out. The Twins pressed together on one side, their shoulders touching, their breathing synchronized. Michael found a corner with the best light and immediately began writing. Curly positioned himself by the door, watching, always watching.

Geoffrey sat in the middle of the floor, too overwhelmed to claim a space, his eyes darting from face to face like he was afraid they might disappear.

"Come on." Peter sat down beside him, close enough that their shoulders touched. "It's all right. We're here."

"I know." Geoffrey's voice was small. "I know we're here. I just… I didn't think it would be like this. I thought—"

"You thought it would be an adventure."

"Didn't you?"

Peter was quiet for a moment. He thought about the island. The real island, the one where he'd been taken as a child. The darkness and the fear and the desperate, clawing need to find the light again.

"I knew it would be dark," he said finally. "I've been to dark places before. But I also know that dark places have windows. You just have to find them."

"How?"

"By looking. By refusing to stop looking, even when everything in you wants to give up." Peter put his hand on Geoffrey's shoulder. "That's what we do, Geoffrey. We're Lost Boys. We find the windows. We find each other. And we don't leave anyone behind.

"Now get some sleep. Tomorrow's going to be—" He hesitated. "Tomorrow's going to be tomorrow. We'll face it when it comes."

Geoffrey nodded. He curled up on his side, using his pack as a pillow, and within minutes, his breathing had steadied into something like sleep.

Peter watched him for a long moment. Sixteen years old. Still growing. Still believing that the world made sense.

He opened his journal and wrote:

First night in the trenches.

The Lost Boys are together. We found a dugout. It smells like death, but it's ours.

Geoffrey asked if I thought this would be an adventure. I don't know what to call it. Adventure seems too small. Nightmare seems too simple. Maybe it's both. Maybe it's neither.

I can hear the guns in the distance. Constant. Like a heartbeat. Like a clock.

Tick. Tick. Tick.

I don't know what that sound is counting down to. I don't think I want to know.

Names so far: The Lost Boys, still alive. The soldier with empty eyes—didn't get his name. The guide who led us here—didn't get his name either.

Tomorrow I'll learn more names. Tomorrow I'll write them down. Tomorrow I'll start building the map of this place, this Neverland, this island of lost boys who might never find their way home.

But tonight, we're still together.

Tonight, that's enough.

Around him, the Lost Boys settled into sleep—fitful, shallow, broken by the distant shells and the closer sounds of rats and the murmuring of men in nearby dugouts who had learned long ago that silence was more frightening than noise.

Peter stayed awake.

He was still awake when the sky began to lighten.

He was still awake when the sergeant came and kicked their boots.

"Stand to," the man said. "Everyone up. Line the wall. Face east."

And Peter Panning, seventeen years old, who had once been stolen and once been found and who would spend the rest of his life somewhere between the two, stood up to face his first dawn in No Man's Land.

***

The dawn came gray and cold.

The Lost Boys stood shoulder to shoulder on the fire step, facing the eastern sky, their rifles clutched in hands that didn't know what to do with them. Around them, the veterans stood the same way—silent, watchful, waiting for something that might or might not come.

"What are we looking for?" Billy whispered.

The soldier next to him didn't turn his head. "Germans. They like to attack at dawn. The light's in our eyes, makes it hard to shoot straight."

"Do they attack every morning?"

"No. Hardly ever, really. But we stand to anyway. Every dawn. Every dusk. Just in case."

"That seems..." Billy hesitated. "Excessive."

"That's the army for you." Now the soldier did turn, and Peter saw his face properly for the first time. He was young—not much older than them, really—with a scar across his chin and eyes that had already learned to be old. "You'll do a lot of excessive things here. You'll stand in mud for hours. You'll eat food that isn't food. You'll sleep sitting up because lying down means drowning. And you'll do it all because the alternative is worse."

"What's the alternative?"

The soldier smiled. It wasn't a happy expression.

"Dying," he said. "The alternative is always dying."

The sun rose. Nothing happened. No Germans came screaming across No Man's Land. No shells fell on their heads. Just light, growing slowly, revealing the wasteland that stretched between the trenches—the wire, the craters, the bodies that hadn't been retrieved.

"Stand down," an officer called. "Breakfast in thirty."

The veterans relaxed, stepped off the fire step, went about the business of being alive for another day. The Lost Boys stayed where they were, staring at the landscape before them.

"Is that—" Geoffrey's voice was faint. "Are those—"

"Yes." Peter didn't look away. "Those are bodies."

They were scattered across No Man's Land like discarded dolls. Some were half-buried in mud, some were draped across wire, some were lying in positions that suggested they'd simply stopped mid-stride and never moved again. The dawn light caught them without mercy, revealing the particular stillness of things that had once been human.

"We have to—someone should—" Billy was struggling to speak. "We can't just leave them there."

"We can. We do." The young soldier from before was still beside them. "Retrieval parties go out at night, sometimes. When it's quiet. But it's never quiet for long, and the parties don't always come back."

"So they just... stay there?"

"Until the mud takes them. Or the shells. Or the rats." The soldier's voice was flat. Matter-of-fact. The voice of someone who had learned to speak about horror without flinching. "Everything disappears eventually. The question is whether anyone remembers it was there."

Peter thought of his journal. Of the names he'd promised to write.

"I'll remember," he said.

The soldier looked at him. Something flickered in his eyes—surprise, maybe. Or recognition.

"Will you?"

"Yes. Every name I learn. Every face I see. I'll write them down."

"Why?"

Peter met his gaze steadily. "Because someone has to. Disappearing isn't the same as being forgotten. And I won't let anyone be forgotten."

The soldier stared at him for a long moment. Then, slowly, he extended his hand.

"Thomas," he said. "Thomas Whitfield. Private. I've been here four months. I've watched a lot of boys make promises they couldn't keep."

Peter shook his hand. "Peter Panning. I keep my promises."

"We'll see." But there was something different in Thomas's voice now. Something that might have been hope. "We'll see."

Peter wrote in his journal:

First dawn. First stand to. First sight of No Man's Land.

The bodies lie where they fall. No one retrieves them. The mud will take them eventually, and no one will remember they were there.

Except me. I'll remember.

New name: Thomas Whitfield. Private. Four months in the trenches. He doesn't believe I'll keep my promise. I'll prove him wrong.

The guns never stopped all night. They're still going now, somewhere in the distance. That rhythm—tick, tick, tick—like a clock counting down.

Counting down to what?

I don't know yet. But I'm going to find out.

First day in Neverland.

Chapter Nine

The Nicknames

France, May 1915

The nicknames were a joke at first. It was Billy who started it—Billy, who couldn't bear silence, who filled every quiet moment with chatter because the alternative was thinking about things that couldn't be thought about. They were huddled in the dugout on their third night, rain drumming on the earth above, water seeping through the timbers and pooling on the floor, and Billy was talking because Billy was always talking.

"You lot need names," he announced. "Proper names. Not the ones your mums gave you."

"What's wrong with our names?" Michael asked.

"Nothing. They're fine names for fine English boys." Billy dug into a tin of bully beef with his fingers. They'd learned quickly that utensils were a luxury the mud didn't allow. "But you're not fine English boys anymore, are you? You're soldiers. You're trench rats. You're..." He waved his hand vaguely. "Whatever we are now. And whatever that is, it needs a different sort of name."

Charlie snorted. "What do you suggest? Mud? Misery? Walking-corpse-who-hasn't-realized-it-yet?"

"Cheerful as always, Slightly." Billy pointed at him with a beef-smeared finger.

"If I'm Slightly, what does that make you?"

"I'm Tootles, obviously."

"Why Tootles?"

"Because I toot." Billy demonstrated with a small, dignified fart. "It's a medical condition. Very serious. The doctors are baffled."

Michael laughed—the first real laugh Peter had heard from any of them since they'd arrived. It sounded strange in the dugout, too bright for the darkness, like a bird that had flown into the wrong room.

"What about me?" Michael asked.

Billy studied him. "Nibs."

"Why Nibs?"

"Because you're always nibbling on something. Biscuits, your pencil, your fingernails when you're nervous. And because you write with a nib." He gestured at Michael's ever-present notebook. "Scribbling away, recording everything. Nibs. It's perfect."

Michael considered this. Then, slowly, he nodded. "Nibs. I can live with that."

"You'll do more than live with it. You'll become it. That's how this works." Billy turned to The Twins, who were pressed together in their corner, watching the proceedings with identical expressions of wary amusement. "Now, you two are tricky. Can't call you by the same name—that's no fun. But you're so alike it's uncanny."

"We're cousins," Thompson said.

"That tends to happen," Thomson added.

"Exactly. So here's what I'm thinking. You're still The Twins. But when I need to tell you apart..." He squinted at them. "Which one of you has the scar?"

"Me." Thompson touched his eyebrow.

"Right. So you're Twin One, and you're Twin Two. Simple. Elegant. Completely unhelpful, which is fitting because you're both completely unhelpful."

"We resent that," Twin One said.

"Deeply," Twin Two agreed.

"But we accept the names."

"Provisionally."

Billy beamed. "Excellent. Now, Curly—"

"Why am I Curly? My name is Curly." The big, quiet boy spoke so rarely that everyone turned to look at him.

"Exactly. That's the joke. You've never let us know your real name. So, it's—"

"It's stupid," Curly said.

"Yes. That too. But you're keeping it."

Curly shrugged. Which, for Curly, was practically enthusiastic agreement. Later, we would find out that his real name was *actually* Curly; named for his mother's paternal grandmother's maiden name.

"Geoffrey," Billy continued, turning to the youngest of them. "You're a problem. You're too young to have a real nickname. You're practically a baby."

"I'm not a baby."

Billy tapped his chin thoughtfully. "I could call you Baby. Or Junior. Or Small One."

"Don't you dare."

"Fine, fine. You're..." Billy's eyes lit up. "You're Keeper. Because you're keeping up with us even though you shouldn't be here. And because—" He glanced at Peter. "Someone told me you're going to be the one who keeps the records someday. When Peter can't."

Geoffrey's face flushed with pride. "Keeper. I like that."

"Good. Because you're stuck with it."

That left Peter.

Billy turned to him, and something shifted in his expression—the joking falling away, replaced by something more serious.

"And you," he said. "You're Pan."

"I know. Ha ha."

"No, I mean..." Billy struggled for words. "You've always been Peter. Peter Panning. But here, in this place, you're something else. You're the one who leads us. The one who keeps us together. The one who—" He gestured at the journal in Peter's hand. "Who makes sure we're remembered."

"That's what Peter Pan does in the story," Michael—Nibs—said quietly. "He leads the Lost Boys."

"Exactly." Billy met Peter's eyes. "So that's who you are now. Pan. Our Pan. The boy who's going to fly us all home."

Peter thought about the story. About Barrie's play, about the boy who never grew up, about the shadow that could be sewn back on and the window that never closed.

He thought about the other Peter Pan—the one who left Wendy behind. The one who forgot. The one who flew away and never came back.

"In the story," he said slowly, "Peter Pan doesn't save the Lost Boys. He just... keeps them. Keeps them on the island forever. They never grow up. They never go home."

"Then we'll write a different story." Billy's gap-toothed grin was fierce. "We'll be the Lost Boys who find their way back. The ones who don't forget. The ones who—"

A shell landed somewhere distant. The dugout shook. Dirt rained from the ceiling.

No one moved.

"The ones who survive," Billy finished quietly. "That's what we'll be. Lost Boys who survive."

Peter looked around the dugout. At Slightly, sardonic and sharp, his mother's fierce love wrapped around him like armor. At Tootles, relentlessly cheerful, Rose's photograph tucked against his heart. At Nibs, scribbling in his notebook, trying to capture something that couldn't be captured. At The Twins, pressed together. At Curly, solid and silent, watching over them all. At Keeper, too young, too eager, burning to prove himself.

His Lost Boys.

His brothers.

And in the darkness of a dugout in France, surrounded by mud and death and the constant *tick-tick-tick* of shells falling somewhere in the distance, the Lost Boys became who they were going to be.

***

The names stuck. Within two weeks, the other soldiers in their sector had picked up the nicknames, shouting for Tootles when they needed someone cheerful, for Nibs when they needed something written down, for The Twins when they needed a job done in perfect synchronization.

It was strange, Peter thought, how a name could change you. How putting on a new identity—even a silly one, even a joke—could make you into someone stronger. The Lost Boys walked taller with their nicknames. Smiled more. Held on to each other more fiercely.

The war was still hell. The trenches were still graves waiting to be filled. But now they had something the other soldiers didn't: a story. A myth. A framework for understanding the incomprehensible.

They were the Lost Boys. They were looking for the window. They were going to fly home.

Even if they didn't believe it—even if part of them knew it was a fairy tale, a comfort, a lie they told themselves to keep moving—they had something to hold on to.

And sometimes, Peter was learning, something to hold on to was the difference between surviving and giving up.

***

Corporal Marsh found them in the second month. He was older than anyone Peter had met in the trenches—forty, maybe, with a face like weathered stone and eyes that had seen too much to be surprised by anything. He wasn't part of their unit; he appeared one morning as if conjured from the mud itself, walking the length of the trench with a notebook in one hand and a careful, cataloguing gaze.

"You're the one who writes names," he said to Peter. It wasn't a question.

Peter looked up from his journal. "Yes."

"May I see?"

Peter hesitated. The journal was private—sacred, almost. The names and stories he'd collected had been entrusted to him, and sharing them felt like a betrayal.

But something in Marsh's eyes made him hand it over.

The corporal read slowly, his lips moving slightly as he traced the entries. *Thomas Whitfield. Private. Four months. Has a sister in Liverpool who sends him socks. Edward Crane. Lance Corporal. Makes shadow puppets at night to help the young ones sleep. James Morrison. Private. Died April 17, 1915. Shell. Quick. Had a dog named Captain.*

"You write about the living and the dead," Marsh said finally.

"Someone has to."

"Yes." Marsh handed the journal back. "Someone does." He studied Peter for a long moment. "How old are you?"

"Eighteen."

"Liar. You're sixteen. Seventeen at most."

Peter didn't answer. Couldn't answer. The lie about his age was the foundation everything else was built on; if it crumbled, he'd be sent home, and his Lost Boys would be alone.

"Relax." Marsh's voice was gentle. "I'm not going to report you. Half the army is children pretending to be men. The other half is men who've forgotten they were ever anything else." He paused. "I do the same work as you, you know. Writing names. Keeping records."

"You're with Graves Registration?"

"Among other things." Marsh sat down on an empty ammunition crate, settling in like he had all the time in the world. "I've been doing this for six months. Walking the trenches. Recording the dead. Making sure they don't just disappear into the mud."

"How many names do you have?"

"In my book? Four hundred and seventeen." Marsh's voice was flat. "In my head? More than I can count. They all blend together after a while. The names become sounds. The faces become shapes. You start to forget that they were people at all."

"I won't forget."

"You say that now. Wait until you've buried your hundredth boy. Your five hundredth. Your thousandth. Wait until the names come faster than you can write them." Marsh met Peter's eyes. "This war is a machine. It takes boys in one end and pushes corpses out the other. And all we can do—people like you and me—is try to make sure the corpses have names."

"That's not enough."

"No. It's not." Marsh stood. "But it's what we have. It's all we have." He reached into his coat and pulled out a small object. It was a watch, Peter saw, old and scratched, its face cracked.

"This belonged to a boy named Daniel," Marsh said. "Daniel Marsh. My son. He was seventeen when he enlisted. Eighteen when he died. Somewhere out there—" He gestured vaguely toward No Man's Land. "He's still lying in the mud. I couldn't find his body. Couldn't bring him home. All I have is this watch and his name in my book."

Peter didn't know what to say. Couldn't find words for the weight of what he'd just been told.

"The watch stopped at 3:47," Marsh continued. "That's when the shell hit. 3:47 in the afternoon, on Christmas Day, 1914. We'd just finished singing carols with the Germans. Did you hear about that? The Christmas truce?"

"Everyone's heard about it."

"Well, it ended at 3:47. For Daniel, anyway. For a lot of boys." Marsh tucked the watch back into his coat. "I keep it with me. Wind it every morning, even though it doesn't work. Reminds me why I'm here."

"Why are you here?"

"To make sure the other Daniels don't disappear. To write their names and remember their faces and tell their mothers what happened to them." Marsh's voice cracked, just slightly. "To do for them what I couldn't do for my son."

He started to walk away, then stopped.

"You're doing good work, Peter Panning. Sacred work. Don't let anyone tell you otherwise." He looked back over his shoulder. "And keep writing. Even when it hurts. Especially when it hurts. That's when it matters most."

He disappeared around a corner of the trench, and Peter was left alone with his journal and the weight of four hundred and seventeen names that weren't his own.

He opened to a fresh page.

Corporal Marsh. Graves Registration. Lost his son Daniel on Christmas Day, 1914. 3:47 p.m. Carries a broken watch that will never tell the right time again.

He says this work is sacred.

I think he might be right.

***

The days developed a rhythm.

Stand to at dawn. Breakfast—if you could call it that—of bully beef and hard biscuits and tea so strong it could strip paint.

Work details: filling sandbags, repairing trenches, hauling supplies through communication lines that seemed designed by someone who had never actually walked through mud. Stand to at dusk. Night watch in shifts. Sleep when you could, where you could, however you could.

And through it all, the guns.

Peter learned to read them the way he'd once learned to read the patterns of light through his mother's prism. The German shells had a particular whistle, a descending scream that told you exactly how much time you had to find cover. The British artillery, responding, had a deeper roar—a sound that was supposed to be comforting but somehow wasn't.

And underneath it all, the constant *tick-tick-tick* of the smaller weapons. Machine guns. Rifles. The endless, mechanical rhythm of killing.

Tick.

A shell landing somewhere.

Tick.

Another, closer.

Tick.

The war counting down to something Peter couldn't name.

"You're listening again," Slightly said one evening. They were sitting in the dugout, playing cards by candlelight, pretending that everything was normal.

"Listening to what?"

"Counting the shells. Moving your lips." Slightly laid down a card. "It's unnerving."

"I'm not counting shells."

"Then what are you counting?"

Peter was quiet for a moment. How to explain it? The feeling that every shell was a name—a life being erased, a story being cut short, a window going dark somewhere that he couldn't see.

"I'm counting the ones I haven't written yet," he said finally. "The names I don't know. The boys I'll never meet because they'll be gone before I can learn their stories."

Slightly stared at him.

"That's morbid, Pan."

"It's true."

"Those aren't the same thing."

"In this place?" Peter looked around the dugout—at Tootles dealing cards, at Nibs scribbling in his corner, at The Twins sitting so close their shadows merged, at Curly watching the door, at Keeper trying desperately to stay awake. "In this place, I think they might be."

A shell landed close. The candle flickered. No one moved.

And the game continued, as games do, until the darkness became too heavy to hold back and they surrendered to something that wasn't quite sleep.

***

Peter dreamed of the island.

The one he'd been taken to as a child, the one with darkness and fear and the desperate, clawing need to escape. This was the other island. The one from Barrie's story. The one with pirates and mermaids and a boy who could fly.

In the dream, he was standing on a cliff overlooking the sea, and the Lost Boys were gathered around him. But they were wrong—twisted, somehow, their faces flickering between the boys he knew and something else. Something hollow. Something dead.

"Pan," they said in unison, their voices layered and strange. "Pan, you promised to take us home."

"I will," he said. "I will, I swear—"

"You can't." The voice came from behind him, and Peter turned to find a figure emerging from the shadows. A man in a long coat, with a hook where his hand should be and eyes that held nothing but darkness. "You can't save them, boy. You can't even save yourself."

"Who are you?"

"You know who I am." The figure smiled—a terrible smile, all teeth and malice. "I'm the thing that follows you. The thing you can't escape. The thing that ticks away the seconds until you run out of time."

Tick.

The figure raised its hook.

Tick.

The Lost Boys began to scream.

Tick.

***

Peter woke gasping. His hands clawed at the dugout floor, seeking purchase, seeking proof that he was alive, that the mud beneath him was real mud and not the nightmare island. His heart hammered against his ribs. His skin was slick with sweat that the cold air turned to ice.

"Pan?" Slightly's voice was groggy with sleep. "Pan, you all right?"

"Fine." The word came out cracked, broken. "Just a dream."

"Same one?"

Peter didn't answer. Didn't need to. Slightly had been woken by his thrashing enough times to know.

"You should talk to someone," Slightly said.

"I'm talking to you."

"I mean someone who can help. A doctor. A chaplain. Someone."

"What would I tell them? That I'm having nightmares about a children's story?" Peter laughed bitterly. "They'd think I was mad."

"Maybe you are. Maybe we all are." Slightly shifted in the darkness. "Pan—the things you saw. Before. When you were a child. The island, or whatever it really was. Do you think this place..." He gestured at the trench walls, the mud, the war. "Do you think it's the same thing? The same kind of darkness?"

Peter was quiet for a long moment.

"I think darkness is darkness," he said finally. "I think it doesn't matter what shape it takes—an island, a war, a man with a hook. It's all the same thing. The thing that wants to swallow you. The thing that counts down the seconds until you stop fighting."

"The crocodile."

"What?"

"In the play. The crocodile that swallowed the clock. It follows Hook everywhere, tick-tick-ticking, counting down to the moment it catches him." Slightly's voice was strange. "Maybe that's what this is. The crocodile. And we're all just waiting for it to catch us."

Tick.

A shell fell somewhere distant.

Tick.

Another, closer.

Tick.

"Then we keep moving. We stay one step ahead," Peter said.

"And when we can't? When it catches us?"

Peter thought of his mother. Of the window. Of the light that scattered but never disappeared.

"Then we make sure someone remembers we were here. We make sure our names are written down." He reached for his journal, held it against his chest like a shield. "We make sure the crocodile doesn't get to swallow us whole."

Slightly was quiet for a long time.

And in the darkness of the dugout, with the guns ticking away the seconds and the crocodile circling somewhere just out of sight, the Lost Boys held on to the only truth they had left:

They were still together.

Chapter Ten

The Ledger of the Lost

France, June 1915

The boy with the wind chimes. We named him Tink.

Peter never learned his real name. None of them did. He'd appeared in their sector three weeks into their deployment—a slight, nervous boy with clever hands and a habit of humming tunelessly under his breath. He'd been assigned to their unit as a replacement for someone who'd been sent home with trench foot, and from the first day, he'd been different.

"What are you making?" Peter asked, watching him twist together pieces of wire and shell casings.

"Wind chimes." Tink held up his creation—a delicate construction of metal fragments, balanced so perfectly that the slightest breeze set them ringing. "I used to make them at home. Before. My mum hung them in the garden. Said they kept the bad spirits away."

"Do they work?"

Tink smiled. It was a strange smile, hopeful and sad at the same time, like he was remembering something beautiful that he knew he'd never see again.

"Maybe," he said. "Maybe not. But they make music. And music's got to count for something, right? Even here?"

He hung the wind chimes from a nail in the dugout ceiling. They caught the air currents that moved through the trench, singing softly. A high, bright sound that had no business existing in a place like this.

"Tink," Tootles said, watching them sway. "That's what you are. Tink. Like Tinkerbell. Making magic out of nothing."

Tink's smile widened. "I like that."

And so Tink joined the Lost Boys—not officially, but in the way that mattered. He sat with them in the dugout. He shared their rations. He learned the rhythm of still here, still together. He made more wind chimes, from anything he could find—shell casings, bullet fragments, bits of wire, pieces of the war transformed into something almost beautiful.

"He's too soft for this place," Slightly said one evening, watching Tink sleep with his latest creation clutched in his hands. "He won't survive."

"None of us are going to survive," Peter said.

"You don't believe that."

"I don't know what I believe anymore." Peter looked at Tink's sleeping face. He was peaceful, young, impossibly innocent. "But I know he deserves to be remembered. If something happens—when something happens—I want to write about the wind chimes. I want someone to know he tried to make music in hell."

"Then write it now." Slightly's voice was soft. "Don't wait until after. Write it now, while he's still here. While you can still see him."

Peter opened his journal.

Somewhere, men are moving pins on maps. Down here, boys are learning how to breathe, quietly. And Tink. Real name unknown. Makes wind chimes from shell casings. Says his mum hung them to keep bad spirits away. He hums while he works—tunelessly, endlessly, like he's forgotten how music is supposed to sound but can't stop trying to remember.

I don't know if the wind chimes keep bad spirits away. But they make the dugout feel less like a grave. They make us feel like something other than corpses waiting to happen.

If he dies—I want the world to know: he made music. In the middle of darkness, he made music. That has to matter. That has to count for something.

***

The gas came on a Tuesday.

They were on a work detail—filling sandbags, the endless, Sisyphean task that consumed half their waking hours—when the sentry started screaming.

"GAS! GAS!"

Peter had practiced this. They all had. The drills in training, the instruction manuals, the officers shouting about procedures and protocols. But training was nothing like the real thing.

The real thing was yellow-green clouds rolling across No Man's Land like fog made of nightmares. The real thing was the smell—a sickly-sweet chlorine stench that burned the inside of your nose before it started on your lungs. The real thing was fumbling for gas masks with hands that shook too badly to work properly, straps tangling, filters not seating, everything wrong, everything too slow.

"MASKS ON! GET YOUR MASKS ON!"

Peter got his mask sealed. Turned to help the others. Slightly was masked. Tootles was masked. The Twins were helping each other. Curly was pulling Keeper's straps tight, his own mask already secure.

Nibs was struggling. His hands were shaking too badly. The straps kept slipping.

"Hold still!" Peter grabbed the mask, forced it over Nibs's face, yanked the straps tight. "Breathe. Slowly. Don't panic."

Nibs's eyes were wild behind the glass eyepieces. But he nodded. He breathed.

And then Peter heard it.

The wind chimes.

Tink was at the far end of the trench, where the sandbag detail had been working. He was standing—just standing—watching the gas roll toward him with something like wonder on his face.

His mask was on the ground. Fallen. Dropped. He was making no move to pick it up.

"TINK!" Peter started running. The gas mask made it hard to breathe, hard to see, hard to move. But he ran anyway. "TINK, YOUR MASK!"

Tink turned to look at him. His face was calm. Peaceful, almost.

"I dropped it," he said. His voice was dreamy, like he was talking about something that had happened to someone else. "When the screaming started. I dropped it, and I can't—I can't find it—"

The gas was twenty feet away. Fifteen. Ten.

Peter dove for the fallen mask. His hands closed around the straps. He turned to throw it—

The gas hit.

It moved faster than he'd expected. Faster than the drills. Faster than anything.

Tink breathed in.

Peter watched his face change. Watched the peacefulness crack and crumble. Watched the boy realize, in the span of a heartbeat, that something terrible had just happened.

"Pan—" Tink's voice was a whisper. "Pan, I can't—"

He fell.

Not dramatically. Not like in the stories. He just... folded. Crumpled. Went down like a puppet whose strings had been cut, his hands clutching at his throat, his mouth opening and closing in silent screams.

Peter dropped beside him. The gas was everywhere now—swirling around them, thick and yellow and hungry. His own mask was sealed, but Tink—Tink was breathing it in with every desperate, choking gasp.

"Hold on." Peter grabbed him, tried to drag him toward clearer air. "Hold on, Tink, I'll get you out—"

But there was no clearer air. There was only the gas, and Tink's face going gray, and the wind chimes singing somewhere overhead like a requiem.

"Wanted..." Tink's voice was barely audible. Bubbling. Wrong. "Wanted to be... heard..."

"You are." Peter's own voice was breaking. "You are, Tink. I hear you. I'll always hear you."

Tink's hand found Peter's. Squeezed once. Weakly.

And then he was gone.

***

They burned the bodies that night.

Not just Tink—there were others, dozens of others, boys who hadn't gotten their masks on in time or whose masks had failed or who had simply been in the wrong place when the cloud rolled through. The gas didn't discriminate. It took everyone it touched.

Peter stood at the edge of the burial pit and watched the flames consume the boy who had made wind chimes.

"You knew him?" Corporal Marsh appeared beside him, his face lit orange by the fire.

"Three weeks. Not long enough to learn his real name."

"But long enough to remember."

"Yes."

Marsh was quiet for a moment. The fire crackled. The smoke rose.

"Write it down," he said finally. "Whatever you know about him. Write it down tonight, while it's still fresh. The details fade faster than you think."

Peter nodded. His hands were shaking too badly to write now, but later—when the flames died down, when the other Lost Boys were sleeping, when the world was quiet enough to hear himself think—he would write.

He would always write.

***

The journal entry was the longest he'd ever done.

Tink. Killed by gas, May 1915, Second Battle of Ypres. He wasn't the first. He won't be the last. But he was ours. He was a Lost Boy. He made music in hell.

I don't know what happens after we die. I don't know if there's a window, a light, something to fly toward. But I know this: as long as these words exist, Tink exists. As long as someone reads his name, he's not forgotten.

In the only way that matters.

The wind chimes keep singing. I think they always will.

Peter closed the journal. Pressed it against his chest. Let himself feel the weight of it—the weight of names, of stories, of lives reduced to ink and paper.

Slightly was watching him from across the dugout.

"You finished?"

"For now."

"Did it help?"

Peter thought about the question. Did writing help? Did trapping someone's life in words make their death any smaller, any more bearable?

"No," he said honestly. "But it's not supposed to help. It's supposed to matter."

Slightly nodded slowly.

Peter looked at the wind chimes, swaying gently in the darkness, singing their high, bright song.

***

In the days that followed, Peter learned the work. Not just the writing—he'd been doing that since he arrived—but the other work. The sacred work, as Marsh called it. The work of finding the dead and giving them back their names.

He started accompanying the burial details. At first, he just watched—watched the men move through the aftermath of battles, identifying bodies, collecting personal effects, marking graves. But soon he was helping. Sorting through pockets for letters, photographs, identification discs. Writing down the details that would let families know what had happened to their sons.

Private Edward Grear. 3rd Battalion. Killed April 23, 1915. Personal effects: one photograph of a woman, identity unknown; one letter, partially legible; one ring, gold, engraved "E.M. to M.C. 1912."

The work was gruesome. The bodies were often unrecognizable—blown apart by shells, decomposed beyond identification, sometimes nothing more than fragments that had to be gathered and labeled "unknown." But Peter did it anyway. He did it because Marsh was right: it mattered. It mattered that someone took the time to look, to search, to refuse the easy anonymity of a mass grave.

"You're good at this," Marsh said one evening. They were walking back from a retrieval, their boots heavy with mud, their hands still stained with the work of the dead. "Most boys can't stomach it."

"I've seen death before."

"Everyone's seen death. That's not the same as doing this." Marsh stopped walking. Looked at Peter with those ancient, grief-carved eyes. "What I mean is—you see them. Not just the bodies. The people they were. The lives they had. Most people can't do that. They look at the dead and see meat. You look at them and see stories."

"My mother taught me." Peter's voice was soft. "She works at a hospital. Great Ormond Street. For children. She used to tell me that every patient was a person first—with a history, a family, a name that mattered. She said the worst thing you could do to someone was forget them."

"Smart woman."

"The smartest. She kept a window open for me, once. When I was lost. She never gave up. Never stopped believing I'd find my way home."

Marsh was quiet for a long moment.

"Is that why you do this? The writing. The names. The stories."

"Part of it." Peter thought of the prism in the nursery window. Of the light that scattered but never disappeared. "Everyone deserves someone who remembers. And if their families can't do it—if they're too far away, too broken, too lost themselves—then someone else has to."

"And that someone is you?"

"That someone is me." Peter met Marsh's eyes. "For as long as I can. For as many names as I can hold."

Marsh reached into his pocket and pulled out the broken watch—the one that had stopped at 3:47, the one that had belonged to his son.

"Daniel would have liked you," he said. "He was the same way. Always writing, always remembering. He used to say that the worst kind of death was being forgotten. That as long as someone spoke your name, you were still alive, somehow, in the only way that mattered."

"He was right."

"Yes. He was." Marsh tucked the watch away. "Keep writing, Peter Panning. And when this war is over—if we survive it—tell people what we did here. Tell them about the burial details and the identification discs and the boys whose bodies were never found. Tell them about the work."

"I will."

And Peter meant it. He would keep writing. He would keep remembering. He would keep the window open for every lost boy who passed through his hands, from now until the end of the war.

From now until forever, if that was what it took.

***

The wind chimes were still singing when Peter finally slept.

He dreamed of Tink—not the Tink who had died gasping in the mud, but the Tink who had hummed tunelessly while he worked, whose hands had transformed destruction into music, who had believed that wind chimes could keep bad spirits away.

In the dream, Tink was flying. Not falling, not dying, not choking on gas—flying. Soaring through a sky that was impossibly blue, his arms spread wide, his face turned toward a light that Peter couldn't see.

And when he woke, the wind chimes were still singing, and the sun was rising over No Man's Land, and somewhere in his chest, a small, fierce hope was burning.

They were going to lose more. He knew that. More names, more stories, more wind chimes falling silent. The crocodile was still circling. The clock was still ticking.

But as long as he had ink in his pen and pages in his book—as long as he could write the names and tell the stories—the darkness wouldn't win.

Not completely.

Not yet.

He opened his journal to a fresh page and began to write.

Chapter Eleven

The Runner of Thunder

France, Summer 1915

The photograph came out every night.

Peter had learned to recognize the ritual: Tootles settling into his corner of the dugout, reaching into his breast pocket, pulling out the small rectangle of card and paper that held Rose's face. He would look at it for exactly three minutes—Peter had timed it—then kiss it once, very gently, and tuck it back against his heart.

Two pats. Always two pats. Like he was reassuring her that he was still there.

"She wrote me again," Tootles said one evening. His gap-toothed grin was softer now than it had been in April, worn down by mud and shells and the particular exhaustion of boys who had learned that survival was not guaranteed. "Want to hear?"

"Always."

Tootles unfolded the letter. It had been read so many times the paper was soft as cloth, the ink faded in places where his fingers had touched it over and over.

"*My darling Billy,*" he read, "*the garden is blooming. The roses—yes, the roses, your favorite joke—are redder than ever this year. Mum says it's because of all the rain, but I think it's because they miss you. Everything misses you. The church bells sound wrong without you there to complain about the early service. The pub is too quiet without your stories. And the couch is too big without you sitting beside me with your arms around me.*"

He paused. His ears went red.

"That's—that's a bit personal. I'll skip that part."

"Probably wise," Slightly said dryly. But his voice was gentle.

"*I count the days until you come home,*" Tootles continued. "*Every morning I mark another number on my calendar, and every night I dream about the day when I can stop marking. When the calendar will just be a calendar again, and you'll be here, and we'll never be apart.*"

His voice cracked.

"*Forever waiting. Forever yours. Your Rose.*"

He folded the letter carefully. Tucked it back into his pocket with the photograph. Patted twice.

"She's got beautiful handwriting," Nibs said. He was the only one who could have said it without it sounding wrong, his writer's eye finding the technical detail in the midst of emotion.

"She's got beautiful everything." Tootles's voice was fierce. "And I'm going to marry her, lads. Soon as this is over. Big wedding, white dress, the whole thing. And you're all invited."

"Even Curly?"

"Especially Curly. He can be the ring bearer. Walk down the aisle looking menacing. Scare the in-laws."

Curly almost smiled. Almost.

"And then what?" Keeper asked. He was young enough to still believe in plans, in futures, in the orderly progression of one thing after another. "After the wedding?"

"House in the countryside. Garden for Rose—she loves gardens. Room for children." Tootles's eyes went distant. "I want six."

"Six children?"

"Six. Three boys, three girls. Enough to make up for..." He stopped. Started again. "Enough to fill a house. Make it loud. Make it alive."

No one said what they were all thinking: *enough to make up for the boys who won't come home. Enough to replace the generation we're burying in the mud.*

But they understood. They all understood.

"You'll get there," Peter said. "You'll marry Rose. You'll have your six children. You'll sit in your garden and tell them stories about the Lost Boys, and they'll think you're making it up, and you'll know it was all real."

"Promise?"

The word hung in the air. The same word they kept asking each other, over and over, as if promises could protect them from shells.

"Still here," Peter said. It wasn't a promise. But it was the closest they could get.

***

Peter became a runner in June.

It happened almost by accident. A message needed to reach the support trenches, the regular runner was down with trench fever, and Peter was standing nearby when the officer started looking for volunteers.

The officer pointed. "You. You look quick. Can you run?"

"I can run."

"Then go. And for God's sake, don't die. I need that message delivered."

Peter ran.

He ran through communication trenches, over duckboards slick with mud, past working parties and burial details and men who looked at him with hollow eyes. He ran through a barrage that erupted halfway through his journey, shells exploding close enough to throw dirt on his helmet. He ran through gas alarms and rifle fire and the constant, ticking awareness that any second could be his last.

He delivered the message.

"You're fast," the officer at the other end said, surprised. "And you're not dead."

"Not yet."

"What's your name?"

"Panning. Peter Panning."

"Like the play?"

"Yes. Like the play."

The officer smiled. It was a tired smile, the smile of a man who had lost too many runners to be surprised by much anymore.

"Pan it is. Welcome to the messenger corps."

***

Running suited Peter. He had always moved toward things rather than away from them—Michael had noticed it back in London, scribbling in his notebook about the boy who faced darkness instead of fleeing. Now he had a use for that instinct. A purpose.

Runners were essential. They carried messages that couldn't be sent by telegraph, orders that needed to reach officers in forward positions, news that might make the difference between victory and disaster. They moved through the most dangerous parts of the battlefield, often alone, always exposed.

Most runners didn't last long.

But Peter was different. He was small and fast and had a preternatural sense for when to duck, when to freeze, when to sprint through a gap in the firing. He learned the communication trenches so well he could navigate them blindfolded. He learned the rhythms of the German artillery, the patterns that told him when it was safe to move.

And he learned to carry more than messages.

On his runs, he started collecting names. The boys he passed in the trenches, the boys in the field hospitals, the boys lying in shell craters waiting for someone to find them. He wrote them all down—quick notes, scribbled in the margins of his ledger, names and details and last words.

Private James Cooper. 2nd Battalion. Wounded in the leg. Said to tell his mother he's sorry about the vase.

Corporal Thomas Reid. Scots Guard. Hit by sniper fire. Asked me to hold his hand while he waited for the stretcher-bearers.

Unknown. Boy. Couldn't have been more than fifteen. Found in crater near section 4. No identification. Blonde hair. Blue eyes. Holding a photograph of a dog.

The ledger grew fat with names. Peter's handwriting grew smaller and smaller, cramming more lives onto each page. He went through one book, then a second, then started on a third.

"You're collecting them," Marsh said one day. They were walking back from a retrieval, another set of dog tags in their pockets, another grave marked in the endless rows. "The boys you meet on your missions."

"Someone has to."

"Most runners don't see them. They just run—eyes forward, feet moving, mind empty. They can't afford to see. If they see, they slow down. If they slow down, they die."

"I see them, and I don't slow down."

Marsh looked at him. "How?"

Peter thought about it. How could he explain the thing that had lived inside him since childhood? The ability to hold horror and hope at the same time, to look at darkness without flinching, to keep moving even when every instinct screamed stop?

"I promised," he said finally. "I promised to remember. And I can't remember if I don't see."

"Most people would break under that weight."

Peter met Marsh's eyes. "Maybe I will too, eventually. But not yet. Not while there are still names to write."

Marsh nodded slowly. His hand went to his pocket, to the watch that would never tell time again.

"You're doing Daniel's work," he said. "You know that?"

"I'm doing my work. It just happens to be the same."

"Maybe that's not an accident."

"Maybe nothing is."

And they walked on, through mud and memory, toward another grave that needed digging.

***

The Twins taught them all the forehead touch. It started as a private thing—their ritual, their reassurance, their way of checking that both halves were still there. But somewhere in the long summer of 1915, it spread.

Thompson did it first with Keeper. The youngest Lost Boy had been caught in a near miss—a shell landing close enough to bury him in dirt, close enough that they'd all thought, for one horrible moment, that he was dead. When they dug him out, gasping and crying and covered in muck, Thompson had grabbed him and pressed their foreheads together.

"You're still here," Thompson said fiercely.

"Still together," Keeper sobbed.

After that, it became theirs. All of theirs.

Tootles did it with Rose's photograph before every mission, pressing his forehead to her paper face like he could feel her through the ink. Nibs did it with his notebook, an absurd gesture that made the others laugh but that meant something, somehow, in the logic of the trenches. Slightly did it with Peter, reluctantly at first, then with a desperation that surprised them both.

"I don't believe in rituals," Slightly said one night, his forehead still warm against Peter's. "I don't believe in magic or prayer or any of that nonsense."

"Then why do you do it?"

"Because belief isn't the point." Slightly pulled back, his sardonic mask firmly in place. "The point is reminding yourself you're still real. Still connected. Still—"

"Here."

"Yes." He almost smiled. "It's stupid. But it works."

"It's not stupid."

"It's extremely stupid. We're grown men pressing our heads together like children."

"We're not grown men. We're boys. And boys need rituals."

Slightly was quiet for a long moment.

"Lilly Ann would like it," he said finally. "The ritual. She's always been one for that sort of thing. Lucky pennies. Salt over the shoulder. Prayers she doesn't believe in but says anyway."

"Tell me about her."

"My mum?" Slightly's voice softened. "She's... impossible to describe. Tiny, fierce, could stare down a charging bull if she had to. She spent years with my father—years of him drinking and shouting and worse—and she never broke. Never stopped fighting. Even when everyone told her to give up, she just... kept going."

"She sounds extraordinary."

"She is. She's the only reason I'm anything at all." Slightly's jaw tightened. "When I told her I was enlisting, she stood in the doorway. Wouldn't move. You remember. Said I'd have to physically move her if I wanted to leave."

"Did you?"

"I picked her up. Set her aside." His voice cracked. "She weighed nothing, Pan. Like a bird. And she just looked at me with those eyes—not crying, not shouting, just *looking*—and she said, 'You come back to me, Charlie Morton. You come back, or I'll find you wherever you are and drag you home by your ear.'"

"She sounds like someone I'd like to meet."

"You would. You'd love her." Slightly met Peter's eyes. "Promise me something."

"Anything."

"If I don't... if something happens... promise you'll tell her. Everything. Not just that I died, but who I was here. What we did. The rituals and the names and the..." He gestured vaguely. "The stupid forehead thing. I want her to know I wasn't just a soldier. I was a Lost Boy."

"You're not going to die."

"Promise anyway."

Peter reached out and pressed his forehead to Slightly's.

"I promise. If anything happens, I'll tell her everything. I'll make sure she knows."

Slightly exhaled slowly. Something in his shoulders loosened.

"Thank you."

"All the way to the end."

"Always."

And in the darkness of the dugout, with the wind chimes singing overhead and the guns ticking in the distance, two Lost Boys held on to each other and pretended that promises could stop the crocodile from ticking closer.

***

The summer stretched on.

Days blurred into nights. Battles blurred into one another—Festubert, Aubers Ridge, names that would become synonyms for slaughter. The Lost Boys ran and fought and dug and buried, and somehow, impossibly, they stayed alive.

Tootles wrote to Rose every day. The letters grew longer as the weeks passed, filled with small details that couldn't possibly survive the censors but somehow did—the color of the mud, the sound of the shells, the way the other Lost Boys looked when they slept.

They're my brothers, he wrote in one letter that Peter glimpsed over his shoulder. *Not by blood—better than blood. I'd die for any of them. I hope I never have to.*

Nibs filled notebook after notebook. His handwriting grew smaller, more cramped, trying to fit more life into each page. He was writing a book, he said—a real book, about all of them. About what it meant to be lost and found and lost again.

"When the war's over, I'm going to publish it. And everyone will know about the Lost Boys. Everyone will know what we did here."

"Will they believe it?" Keeper asked.

"They won't have to believe it. They'll feel it." Nibs touched his notebook like it was sacred. "That's what stories do. They don't make you believe things—they make you *feel* things. And once you feel them, they become true."

Curly stood watch. He was always standing watch, always scanning for threats, always positioning himself between the younger boys and the darkness. He still wrote to Emma every day—short letters, practical letters, nothing like Tootles's romantic outpourings—but Peter saw the way his face softened when he read her replies.

She's learning to read, Curly said once. *Properly, I mean. She's only ten, but she's doing it. Teaching herself, because I'm not there to help.*

"She'll be a scholar by the time you get back," Peter said.

"She'll be anything she wants to be." Curly's voice was fierce. "I'm going to make sure of it."

And The Twins stayed together. Always together. Their ritual evolved—not just foreheads touching, but hands clasped, shoulders bumping, an endless series of small contacts that reminded them they were still two halves of one whole.

***

On the last day of August, Peter ran a message through a barrage that killed eleven men.

He didn't know their names. He was running too fast, too focused, too determined to deliver the order that might save a hundred more. He heard the shells falling, felt the earth shake, saw bodies blown apart in his peripheral vision—but he kept running.

When he reached his destination, he was shaking so badly he could barely hand over the message.

"Good work," the officer said. He didn't even look up from the paper.

Peter stumbled out of the command post and vomited against a wall.

He stayed there for a long time, forehead pressed against the sandbags, stomach heaving, mind spinning with the faces he hadn't seen, the names he hadn't learned, the eleven men who had died while he ran past them without stopping.

That night, in the dugout, he opened his ledger and wrote:

Eleven men. August 31, 1915. Killed in barrage near section 7.

I don't know their names. I was running. I couldn't stop. If I'd stopped, I would have died too, and the message wouldn't have been delivered, and more men would have died in the attack it was meant to prevent.

Marsh told me once that I'd have to make choices. That the work—the sacred work of remembering—couldn't always come first. That sometimes I'd have to choose between saving the living and honoring the dead.

Today I chose the living.

I don't know if it was right. I don't know if "right" means anything here. All I know is that there are eleven men who will be buried without anyone writing their stories. Eleven names that will never make it into this ledger.

And I'll carry that. I'll carry all of them—the names I wrote and the names I didn't, the boys I remembered and the boys I forgot. I'll carry them until the war is over, and then I'll carry them longer, and then I'll carry them forever.

That's the work.

That's the window I keep open.

That's the only thing I know how to do.

He closed the ledger and pressed it against his chest.

Somewhere above him, Tink's wind chimes sang their high, bright song.

Chapter Twelve

The Truce That Wasn't

France, December 1915

The snow began falling on Christmas Eve. It came softly at first—a few flakes drifting down from a sky the color of old pewter, melting as soon as they touched the mud. But as the afternoon faded into evening, the snow thickened, layering itself over the trenches like a blanket, like a shroud, like a world trying to remember what peace looked like.

Peter watched it fall from his spot on the fire step.

The guns had been quiet for hours—not silent, never truly silent, but subdued. A shell here and there, a rattle of machine gun fire in the distance, but nothing like the constant hammering of ordinary days. Even the war, it seemed, was taking a breath.

"Beautiful, isn't it?"

Marsh appeared beside him, his face turned up to the sky. Snowflakes caught in his graying hair, on his eyelashes, on the shoulders of his mud-caked coat.

"I suppose." Peter couldn't quite feel the beauty. The snow was just another form of cold, another way for the weather to make their lives miserable. "It'll turn to slush by morning. Mix with the mud. Make everything worse."

Marsh didn't seem bothered by this. "Probably. But right now, in this moment, it's beautiful. Sometimes that has to be enough."

Peter looked at him—at this man who had buried hundreds of boys, who carried the weight of a thousand names, who had somehow learned to find beauty in the spaces between horrors.

"You're thinking about last year," Peter said. "Christmas 1914."

"I'm always thinking about it. This time of year especially." Marsh pulled out his watch—the broken watch, the one frozen at 3:47—and held it in his palm. "I keep hoping it will happen again. That the singing will start, and the boys will climb out, and for one day—just one day—we'll remember we're all human."

"The officers won't allow it. Not this year."

"No. They've made that very clear." Marsh's voice was bitter. "Fraternization with the enemy is forbidden. Anyone caught attempting to repeat last year's events will face court martial."

"So we just... pretend it never happened? Pretend that we didn't play football with the people we're supposed to kill?"

"That's what they want. The generals. The politicians. The men who write the rules from offices a hundred miles from the nearest shell." Marsh closed his hand around the watch. "They're afraid, you see. Afraid of what happens when soldiers realize the enemy has faces. Names. Families waiting for them to come home."

"They're afraid we'll stop fighting."

Marsh tucked the watch away. "They're afraid we'll start thinking. That's far more dangerous."

***

The singing started at midnight. Peter was half-asleep in the dugout, wedged between Slightly and Tootles, when he heard it. It was faint at first, barely audible over the wind, then grew stronger. A melody he recognized. Words he didn't understand.

Stille Nacht, heilige Nacht...

"Is that—" Tootles sat up, his eyes wide. "Is that *German*?"

"'Silent Night.'" Nibs was already reaching for his notebook. "They're singing 'Silent Night.' In German."

The Lost Boys scrambled out of the dugout, climbing up to the fire step, pressing against the parapet to hear. All along the trench, other soldiers were doing the same—heads appearing above the sandbags, faces turned toward the darkness where the enemy waited.

Alles schläft, einsam wacht...

The voices were young. That was what struck Peter first. Young voices, carrying across No Man's Land, singing a carol about peace and holiness in a place that had known neither for eighteen months.

"They're just boys," Keeper whispered. "Like us."

"Of course they are." Slightly's voice was strange. Tight. "Did you think they were monsters?"

"I don't know what I thought. I never..." Keeper swallowed. "I never thought about them at all. They were just... the enemy. Shapes in the distance. Things to shoot at."

"That's how they want you to think. Makes it easier." Slightly's jaw was clenched. "But listen to them, Keeper. Really listen.

They're singing the same song we sang in church back home. Their mothers taught them the words, just like ours did. They're cold and scared and missing their families, just like us."

Schlafe in himmlischer Ruh...

The song ended. Silence fell—not the suffocating silence of waiting for shells, but something gentler. A held breath. A pause in the relentless *tick-tick-tick* of the war.

And then, from somewhere in the British trench, a single voice began to sing.

Silent night, holy night...

Peter didn't see who started it. The voice just appeared—a tenor, shaky but clear, picking up the melody in English. Other voices joined. Then more. The song swelled, rising from both trenches, German and English twining together in the cold air.

Two armies, two languages, one melody.

Peter felt something crack in his chest.

***

"It's not the same as last year," Marsh said quietly. They were standing together on the fire step, watching the snow fall, listening to the carols that passed back and forth across No Man's Land. "There's no climbing out. No meeting in the middle. The officers have made the consequences too clear."

"But they're still singing."

"Yes. They're still singing." Marsh's voice held a note Peter couldn't place. "That's what gives me hope, you know. Not the truces—those can be forbidden, punished, stamped out. But the singing? The impulse to make music in the darkness? That can't be taken from us. No matter how hard they try."

A new song started on the German side. One Peter didn't recognize—a folk tune, maybe, or a hymn from a tradition he didn't know. But it was beautiful. Hauntingly, impossibly beautiful.

"I'm going out," Peter said.

Marsh turned sharply. "What?"

"I'm going out. Into No Man's Land. Just for a moment."

"Are you mad? The officers—"

"The officers aren't here." Peter was already moving toward the ladder. "And I need to—I don't know. I need to see. I need to know if there's anything left of what you described. The humanity. The connection."

Marsh grabbed his arm. "Peter. If they catch you—"

"Then they catch me." Peter met his eyes. "But I can't just stand here and listen. I can't just watch from a distance while they sing about peace. I have to *do* something."

"What can you possibly do?"

"I don't know yet. But I have to try."

Marsh stared at him for a long moment. Then, slowly, he released Peter's arm.

"Take this." He pressed something into Peter's hand—the broken watch, stopped at 3:47. "If something happens to you out there—if you don't come back—I want that watch to go home with a story. I want someone to know what it meant."

"Marsh—"

"Go. Before I change my mind." His voice was rough. "And for God's sake, be careful."

***

No Man's Land was different in the snow. The craters were softened, their edges blurred to gentle curves. The wire caught the moonlight, sparkling like twisted tinsel. Even the bodies—the ones that hadn't been retrieved, the ones that had become part of the landscape—were hidden beneath a blanket of white, their shapes reduced to gentle mounds.

Peter moved slowly, carefully, his breath clouding in the cold air. The singing continued—German carols, English carols, sometimes both at once, harmonies emerging from chaos the way they always did when human voices reached for each other across distances.

He stopped in the middle of the wasteland. Stood there, exposed, vulnerable, a single figure in an ocean of white.

And he waited.

The German voice called out first. "Englishman! You are alone?"

Peter's German was limited—a few phrases picked up from prisoners, from the dying boys he'd tended on both sides of the line. But he understood the question.

"Alone," he called back. "Just me."

A pause. Then:

"Why you come out?"

"To listen. To see if..." Peter struggled for words. "To see if we're still human."

Laughter. Not cruel—surprised, almost delighted. Then the voice again, closer now:

"You want to know if we are human? Come. I show you."

A figure emerged from the darkness. A German soldier, young—Peter's age, maybe, or close to it—with a round face and nervous eyes and hands held carefully visible.

"I am Wilhelm," the soldier said. His English was better than Peter's German. "My friends call me Willy."

"Peter. My friends call me Pan."

Wilhelm smiled. "Pan? Like the forest god? Strange name for a soldier."

"Strange war for any of us."

They stood face to face in the snow, two boys who should have been killing each other, speaking instead about names and music and the incomprehensible fact of their shared humanity.

"You like the singing?" Wilhelm asked.

"It's beautiful."

"My mother taught me. These songs. She sing them every Christmas, since I am small." Wilhelm's voice softened. "I think of her now. What she is doing. If she lights the candles, says the prayers, cries for me."

"I think of mine too." Peter touched the pocket where his journal rested. "She keeps a window open. Back in London. Says she'll never close it until I come home."

"A window?"

"For the light. So I can find my way back."

Wilhelm nodded slowly. Something passed between them—an understanding that went beyond language, beyond nations, beyond all the forces that were supposed to make them enemies.

"I have something," Wilhelm said. He reached into his coat, and Peter tensed—an automatic response, trained into him by months of war—but what Wilhelm produced was not a weapon. It was a photograph.

"My girl," Wilhelm said, holding it out. "Anna. We are to be married. After."

Peter looked at the photograph. A young woman, dark-haired, smiling shyly at the camera. She looked remarkably like Rose.

"One of my friends has a girl too," Peter said. "Rose. He keeps her photograph next to his heart."

"Then we are the same." Wilhelm tucked Anna's picture away. "We fight for them. The girls who wait. The mothers who pray. The homes we hope to see again."

"Do you think we will? See them again?"

Wilhelm was quiet for a long moment. The snow fell around them, silent and indifferent.

"I think some of us will," he said finally. "And some of us will not. And the ones who do not..." He touched his chest, over his heart. "They will live here. In the ones who remember."

***

They talked until the sky began to lighten. Peter learned that Wilhelm was a baker's son from Munich. That he had enlisted because his friends had enlisted, because the kaiser had called, because saying no had never occurred to him. That he had been in the trenches since August 1914, had survived battles Peter had only read about in newspapers, had buried friends whose names he still whispered every night before sleeping.

"We are not so different," Wilhelm said. "Your army and mine. We both think we are right. We both think we fight for something good. And we both..." He gestured at the wasteland around them. "We both end up here. In the mud. In the dark. Wondering how this happened."

"How did it happen?"

"I do not know. The old men decided. The kaisers and kings and politicians. They drew lines on maps and sent us to die on them. And we went because we believed them, because we trusted them, because we did not know what else to do."

"And now?"

Wilhelm smiled sadly. "Now we know. Now we have seen what war truly is. But it is too late. The machine is running. We cannot stop it. We can only..." He searched for the word. "We can only survive. And remember. And hope that someday, someone will listen when we tell them what we saw."

"I'll listen," Peter said. "I'll remember."

"I know." Wilhelm reached out and clasped Peter's hand. "I can see it. In your eyes. You are one who carries. One who does not forget."

"Neither are you."

"Perhaps. But I do not have—" Wilhelm gestured at Peter's coat, where the journal made a slight bulge. "I do not have words. You do. Use them. Tell our story. All of it. The horror and the humanity. Tell them that on Christmas night, two boys met in the middle of death and found each other still alive."

"I will."

"Promise?"

Then, in careful German, Peter said: "Noch hier. Noch zusammen."

Wilhelm's eyes widened. Then he laughed—a real laugh, surprised and delighted.

"Your German is terrible," he said.

"I know."

"But I understand." Wilhelm squeezed Peter's hand once more, then released it. "Go. Before the light comes. Before someone sees."

"Will I see you again?"

"I do not know. Perhaps. In this war, all things are possible." Wilhelm began to back away toward his own lines. "But if I do not—if the machine takes me—remember. Remember Wilhelm. Remember Anna. Remember that we were here."

"I'll remember."

"Fröhliche Weihnachten, Peter Pan."

"Merry Christmas, Willy."

And then Wilhelm was gone, swallowed by the darkness and the snow, leaving Peter alone in No Man's Land with the dawn approaching and the weight of another name settling into his heart.

***

Marsh was waiting when Peter climbed back into the trench.

"Well?"

"I met someone." Peter's voice was strange in his own ears. Distant. "A German boy. Wilhelm. He has a girl named Anna. He wants to be a baker after the war."

Marsh nodded slowly. "And?"

"And we talked. For hours. About the war and our families and what it means to be human in a place like this." Peter reached into his pocket and pulled out the broken watch. "I didn't need this. He didn't try to kill me. He just... talked to me. Like I was a person."

"Because you are."

"Because we both are." Peter pressed the watch into Marsh's hands. "Keep it. Keep carrying Daniel with you. But know that somewhere on the other side..." He gestured toward the German lines. "There are boys just like him. Just like us. Boys who want to go home."

Marsh was quiet for a long moment. The snow was beginning to melt now, the brief beauty fading back into mud and gray.

"Write it down," he said finally. "What happened tonight. What Wilhelm told you. All of it."

"I will."

"And Peter?"

"Yes?"

"Merry Christmas."

Peter looked at the trench—at the mud and the sandbags and the boys who were starting to stir in their dugouts, waking to another day of war. At the wind chimes that still sang in the entrance to their shelter, Tink's legacy, still making music in hell.

"Merry Christmas, Marsh."

***

The dugout was warm when Peter entered—warm with bodies and breath and the particular closeness of boys who had learned to sleep in piles. The Lost Boys were awake, gathered around a single candle, their faces turned toward something Tootles held.

Keeper spotted him first. “Pan! Where have you been? We woke up and you were gone—”

“Out.” Peter sat down heavily. His legs were shaking. Exhaustion was finally catching up with him, the adrenaline fading. “I went out. To No Man's Land.”

Silence. The Lost Boys stared at him.

“You did *what*?” Slightly's voice was sharp with fear. “Pan, do you have any idea—the officers would have—if anyone had seen you—”

“I met a German.”

The silence deepened.

“His name is Wilhelm. He wants to be a baker. He has a girl named Anna.” Peter looked at Tootles. “She looks like Rose. He keeps her photograph over his heart, just like you.”

Tootles's hand went automatically to his breast pocket. Patted twice.

“You talked to him?” Nibs asked. His notebook was already out. “You actually talked to a German soldier?”

“For hours. About everything. He's not a monster, Nibs. He's just a boy. Just like us. Lost in the same war, looking for the same window.”

Peter looked around at his Lost Boys—his brothers, his family, the boys he would die to protect. “That's what I learned tonight. We're all still here. We;re all still together. British and German, Allied and enemy. We're all just boys, trying to find our way home.”

“That's beautiful,” Nibs said. He was writing furiously. “That's... Pan, that's exactly what my book needs. That's the truth of it.”

“Write it down. All of it.” Peter's eyes were closing. The exhaustion was overwhelming now. “Write about Wilhelm. Write about the singing. Write about two boys meeting in the middle of death and finding each other still human."

"I will."

"And if I don't survive—if any of us don't survive—make sure the story does. Make sure someone knows that even here, even in the worst place in the world, we never stopped being human."

"Promise. Now sleep, you reckless idiot. You've earned it."

Peter slept.

And if he dreamed of a German boy with Anna's photograph, of a broken watch stopped at 3:47, of songs rising from two trenches and meeting in the snow—well.

Some dreams were worth remembering.

Some dreams were worth writing down.

***

In his journal, later, Peter wrote:

Christmas Day, 1915.

They tried to stop it. The generals, the officers, the men who make the rules. They forbade fraternization. They threatened courts martial. They did everything they could to keep us apart.

But we sang anyway.

Across No Man's Land, across the wire, across all the boundaries they drew to divide us—we sang. The same songs. The same melodies. The same prayers in different languages.

And in the middle of the wasteland, I met a boy named Wilhelm. He told me about Anna. About Munich. About his mother's Christmas traditions and his dreams of baking bread in a shop of his own someday.

He's not my enemy. I don't know if he ever was. He's just another lost boy, looking for a window, hoping the light hasn't gone out.

Fröhliche Weihnachten, Wilhelm. Merry Christmas.

Chapter Thirteen

The Hook

France, Spring 1916

The nightmares grew worse after Christmas.

Peter had hoped that meeting Wilhelm might quiet something inside him—might prove that humanity could survive in the trenches, that the darkness didn't have to win. And in his waking hours, it did help. He carried Wilhelm's name like a talisman, wrote about their meeting in letters home, told the story to anyone who would listen.

But at night, the crocodile still circled.

The dream was always the same. He stood in No Man's Land, but it wasn't No Man's Land—not quite. The mud was there, and the wire, and the distant shapes of trenches cutting through the earth like wounds. But the sky was wrong. Too dark. Too close. Pressing down like a lid on a coffin.

And the silence.

That was the worst part. The guns had stopped. The shells had stopped. Even the wind had stopped, leaving nothing but a stillness so complete it felt like suffocation.

Peter tried to move, but his feet were sunk in the mud, held fast by something that felt like hands. He looked down and saw fingers, dozens of fingers, gray and rotting, reaching up from the earth to grip his ankles.

The dead.

The dead were holding him.

He tried to scream, but no sound came out. His throat was locked, his voice stolen, his whole body frozen in a terror that went beyond anything he'd felt in waking life.

And then he heard it.

Tick.

Distant. Muffled. Like a heartbeat heard through water.

Tick.

Closer now. The mud began to tremble.

Tick.

The fingers tightened around his ankles. The dead were pulling him down, dragging him into the ground, and he couldn't fight them, couldn't run, couldn't do anything but watch as the darkness rose to swallow him whole.

And then the voice.

"Hello, boy."

***

Peter woke gasping, clawing at the dugout floor, his whole body drenched in sweat that the cold air turned to ice.

"Same one?" Slightly's voice came from the darkness. They'd stopped asking if he was all right. The answer was always the same, and they all knew it was a lie.

"Same one." Peter pressed his palms against his eyes, trying to push the images away. "The mud. The hands. The—"

"The voice?"

Peter nodded. He'd told Slightly about the voice—about the figure that emerged from the darkness in every dream, the man with the hook where his hand should be, the eyes that held nothing but shadow.

"Have you thought about what I said?" Slightly asked. "About talking to someone?"

"Who would I talk to? The chaplain? He'll tell me to pray. The medics? They'll say I'm shirking." Peter laughed bitterly. "The dreams aren't the problem, Slightly. The war is the problem. The dreams are just... the war following me into sleep."

"That's exactly what shell shock is."

"I don't have shell shock."

"Pan—"

"I *don't*." Peter's voice was fiercer than he intended. "I'm not broken. I'm still fighting. I'm still running. I'm still writing the names."

"None of those things mean you're not broken."

"Then what does being broken mean?"

Slightly was quiet for a long moment. When he spoke again, his voice was soft.

"My father was broken. Back in London. He drank because he was broken, and he hit us because he was broken, and he never got better because he never admitted anything was wrong." Slightly shifted in the darkness. "I'm not saying you're like him. I'm saying that ignoring a wound doesn't make it heal. It just makes it fester."

"What do you want me to do? Stop dreaming?"

"I want you to let us help. The Lost Boys. We're your brothers, Pan. We're supposed to carry each other." Slightly reached out to Peter. "That means the dark parts too."

Peter closed his eyes. Her let himself feel the warmth, the reminder that he wasn't alone.

"There's a figure in the dream," he said quietly. "A man. He has a hook for a hand."

"A hook?"

"Like Captain Hook. From Barrie's play." Peter pulled back, opening his eyes. "I know how it sounds. I know it's absurd—dreaming about pirates when I'm surrounded by real horrors. But he's there, every time. Standing in the darkness, waiting for me."

"What does he say?"

"'Hello, boy.'" Peter's voice dropped. "And then he says other things. Things I can't remember when I wake up. But they feel true. They feel like..." He struggled for words. "Like he's telling me something I already know. Something I don't want to admit."

"What kind of something?"

"That I can't save them. Any of them. That the war is going to take everyone I love, and there's nothing I can do but watch." Peter's hands were shaking. "That I'm going to become him. Become Hook. Become the thing I'm fighting against."

"Pan—"

"He's not the enemy, Slightly. That's what the dream is trying to tell me. Hook isn't the Germans. Hook isn't even the generals. Hook is what the war does to us. The damage. The darkness. The part of ourselves that gets twisted by all this death until we don't recognize it anymore."

Slightly was silent for a long time.

"Then we fight him," he said finally. "The same way we fight everything else. Together."

"How do you fight something inside yourself?"

"I don't know. But we'll figure it out." Slightly's voice was fierce. "You're not becoming Hook, Pan. You're not losing yourself. I won't let you."

"What if you can't stop it?"

"Then I'll stand beside you while it happens. And I'll remind you who you really are until you believe it again."

Peter felt something break inside him. Not the darkness—that was still there, and it always would be—but the wall he'd built around it. The pretense that he was fine, that he was strong, that he could carry everything alone.

Slightly pressed their foreheads together again. "Now go back to sleep. And if Hook comes for you, tell him he'll have to go through me first."

Peter almost smiled. "You'd fight Captain Hook?"

"I'd fight anyone for you. Even imaginary pirates." Slightly's voice held a warmth that cut through the cold. "That's what brothers do."

***

Days blurred into weeks, weeks into months. Stand to at dawn, work details, ration runs, night watches, all of it cycling endlessly like the hands of a clock that never ran down.

Tick. Tick. Tick.

Peter ran messages. He collected names. He wrote in his ledger every night, recording the living and the dead, building a catalogue of souls that grew thicker with each passing day.

And he dreamed of Hook.

The dreams didn't stop. They evolved, deepened, became more complex. Sometimes Hook spoke to him about the war—about the futility of it, the endless grinding machine that turned boys into corpses. Sometimes Hook showed him things: visions of the future, of battles yet to come, of friends dying in mud that hadn't been churned yet.

Sometimes Hook just stood there, in the darkness, waiting.

"You'll understand soon," he said in one dream. "You'll see what I am. What I always was."

"I'll never be you."

"You already are, boy. Every name you write, every death you witness, every piece of yourself you bury to keep moving—that's me. Growing. Waiting." Hook smiled his terrible smile. "The question isn't whether you become me. The question is whether anyone will be left to notice."

Peter woke screaming.

***

"Tell me about Hook," Nibs said.

They were on a work detail—filling sandbags, the eternal task—and Nibs had his notebook out, scribbling between shovelfuls of mud.

"What do you want to know?"

"Everything. The dreams, what he says, what he looks like." Nibs's eyes were bright with the intensity he got when he was chasing a story. "I'm writing about all of us, Pan. The whole truth. And if Hook is part of your truth, I need to understand him."

Peter drove his shovel into the mud. Lifted. Dumped. Again.

"He's not a person," he said finally. "He's a... a symbol, I suppose. Of what the war does to us. The damage it leaves behind."

"But he looks like the Hook from Barrie's play?"

"Yes and no. He has the hook, the coat, the theatrical menace. But his face changes. Sometimes he looks like the officers who send us over the top. Sometimes he looks like the politicians in the newspapers. Sometimes..." Peter paused. "Sometimes he looks like me."

Nibs stopped writing. "Like you?"

"An older version. Harder. Emptier." Peter's voice was distant. "A version that's survived so long he's forgotten what he was surviving for."

"That's not going to be you, Pan."

"How do you know?"

"Because you write." Nibs held up his notebook. "Same as me. You put everything into words, and words don't forget. As long as you keep writing, you keep remembering. And as long as you keep remembering, you stay you."

"What if the words aren't enough?"

"Then we write more. We write until our hands cramp and our eyes blur and we can't hold the pen anymore." Nibs's voice was fierce. "And then we pass the pen to someone else. Keeper. The Twins. Whoever's left. Because the words are bigger than any of us. The story keeps going even when the storytellers don't."

Peter looked at his friend—this boy who had once been Michael Davies from London, who scribbled in notebooks and dreamed of writing books, who still believed that stories could save them.

"You really think writing can stop Hook?"

"I think writing is how we prove Hook didn't win. Every name in your ledger, every page in my notebook—that's evidence. Proof that we were here, that we mattered, that the war didn't erase us completely."

"What if no one reads it? After?"

"Someone will. Someone always does." Nibs smiled. "That's the magic of words, Pan. They wait. They're patient. They sit on shelves and in boxes and at the bottom of trunks, and they wait for the person who needs them. Maybe it takes ten years. Maybe a hundred. But eventually, someone opens the book, and the story comes alive again."

Peter thought about his mother's prism. About light scattering, finding new places to land.

Nibs returned to his notebook. "Now tell me more about Hook. I want to get him right."

***

The letter from Rose arrived in March. Tootles read it in the dugout, his face cycling through emotions too fast to track—joy, then fear, then something that looked like wonder.

"What is it?" Slightly asked. "What did she say?"

"She says—" Tootles's voice cracked. "She says she's counting the days. She says she dreams about our wedding. She says..." He held up the letter, his hands shaking. "She says she loves me. Actually wrote it out. 'I love you, Billy Hawkins, and I always will.'"

"She's never written that before?"

"Not like this. Not so... direct." Tootles pressed the letter to the chest pocket where Rose's photograph lived. "It's like she knows something. Like she's trying to tell me while she still can."

A chill ran through the dugout.

"Don't talk like that," Curly said. It was more words than he'd spoken in days. "Don't borrow trouble."

"I'm not. I'm just..." Tootles looked around at them, at his Lost Boys, his brothers. "Something's coming. I can feel it. Everyone can feel it. The war's building toward something big. And I want—I need—" He swallowed. "I need you all to know. If something happens to me. Rose is real. Our love is real. It's not just a story I tell myself to get through the days."

"We know," Peter said quietly.

"Do you? Do you really?" Tootles pulled out the photograph. Rose smiled up at them, frozen forever in a moment of joy. "This girl. This girl right here. She's waiting for me. She's going to be my wife. We're going to have children and a house and a life, a real life, and I'm going to grow old with her. That's the future I'm fighting for. That's why I'm still here."

"Still together," The Twins said in unison.

Tootles kissed the photograph. Two pats. "Still together. And when this war is over—when we're all home, all of us—I'm going to invite you to the wedding. Every single one of you. And Rose will meet the Lost Boys, and she'll understand why I survived. Because of you. Because we carried each other."

"It's a date," Slightly said. His voice was rough. "I'll wear my best suit."

"Me too," Keeper added.

"We'll be there," The Twins said.

Curly nodded.

Nibs was writing furiously.

And Peter—Peter was thinking about Hook. About the shadow that waited in his dreams, counting down the seconds until it could swallow them all.

Something's coming.

Tootles was right. Everyone could feel it. The buildup of artillery, the increase in patrols, the officers walking around with maps and grim expressions. The war machine was winding up for something massive. Something that would change everything.

"Write to her," Peter said. "Tonight. Tell her everything you just told us."

"The censors—"

"Find a way. Write in code. Write in metaphor. Just make sure she knows." Peter's voice was urgent. "Make sure she knows you love her. Make sure the words exist somewhere, even if—"

He stopped. Couldn't finish. Tootles understood anyway.

"I will." He tucked Rose's photograph back into his pocket. Patted twice. "I'll write the best letter I've ever written. And when she reads it, she'll feel me there with her. Even if I'm a thousand miles away."

"Even if you're farther than that," Slightly murmured.

No one asked what he meant.

They all knew.

***

Peter wrote in his ledger that night:

March 1916.

Tootles got a letter from Rose. She told him she loves him. Actually wrote the words, for the first time. He cried—we all pretended not to notice—and then he made us promise to come to his wedding.

I made the promise. We all did.

But Hook was in my dream again last night. Standing over Tootles. Looking at Rose's photograph with those empty eyes.

"You can't save him," Hook said. "You can't save any of them. The crocodile is coming, and it's hungry, and it doesn't care about love letters or wedding plans or all the pretty futures you've built in your heads."

I told him to go to hell.

He laughed.

"Where do you think we are, boy? Where do you think we've been this whole time?"

I woke up before he could answer. Before I had to see what he meant.

Something's coming. I can feel it. The war is building toward something terrible.

All I can do is keep writing. Keep remembering. Keep the window open for as long as I can.

Chapter Fourteen

The First Day

France, July 1, 1916

The bombardment had been going for seven days.

Seven days of shells falling on the German lines. Seven days of explosions so constant they stopped being sound and became a kind of weather—something you lived inside, something that pressed against your skin and rattled your bones and made every thought feel like it was being shouted into a hurricane.

The officers said it would destroy everything. The wire, the trenches, the machine guns. By the time the infantry went over, they said, there would be nothing left but craters and corpses.

"A cakewalk," one captain called it. "The Germans will be dead or running. All you have to do is walk across and plant the flag."

Peter watched the shells fall and tried to believe it.

***

The night before the assault, Tootles took out Rose's photograph.

It was worn now—softer at the edges, creased where he'd folded it wrong, the image fading from the oils of his fingers. But Rose still smiled from within the frame, still laughed at something beyond the camera, still looked like summer and hope and everything the war wasn't.

"I wrote her," Tootles said. The Lost Boys were gathered in the dugout, pressed together, drawing comfort from each other's warmth. "Last night. Told her everything I wanted to say. Just in case."

"Just in case," Slightly echoed. His voice was hollow.

"I told her about the wedding. Where I want to hold it—the church in her village, the one with the roses in the garden. I told her about the house I'm going to build us—three bedrooms upstairs, kitchen with a big window, garden out back for her flowers." Tootles's voice cracked. "I told her about the children. Their names. I've got them all planned out."

"What names?" Keeper asked quietly.

"William for a boy. After my dad. Margaret for a girl. After her mother." Tootles stroked the photograph. "And if we have more, we'll name them after all of you. Charlie and Michael and Peter and—"

"Don't." Slightly's voice was sharp. "Don't do that. Don't name your imaginary children after us like we're already dead."

"That's not what I—"

"That's exactly what you mean. You're saying goodbye. You're putting everything in order, tying everything up, making sure you've said all the things that need saying." Slightly's eyes were bright with something that might have been anger and might have been tears. "Well, stop it. We're not dying tomorrow. None of us.

We're going to walk across that field and plant the bloody flag, and then we're going to come back and you're going to marry Rose and have all those children and I'm going to dance at your wedding. Do you understand?"

Tootles stared at him.

"Charlie—"

"Do you understand?"

A long pause. The bombardment roared overhead, shaking dirt from the ceiling.

"I understand," Tootles said softly.

"Good." Slightly turned away. His shoulders were shaking.

Peter watched them both—the optimist and the cynic, the boy who believed in after and the boy who refused to say goodbye—and felt the weight of everything he couldn't say pressing against his chest.

The Twins pressed their foreheads together. Curly was watching the door, as always, as if he could guard them against the morning. Keeper had his eyes closed, lips moving in what might have been prayer. Nibs was writing—still writing, always writing, recording every moment until there were no moments left.

And Tootles was looking at Rose.

"I can hear her laugh," he said quietly. "Sometimes. When everything's quiet. I can hear her laughing, like she's right here with me." He pressed the photograph to his lips. "I'm going to hear it again. For real. I'm going to make her laugh every day for the rest of our lives."

"You will," Peter said. "I promise."

But even as he said it, Hook was there—in the corner of his vision, in the shadows of the dugout, watching with those empty eyes.

You can't save him. You can't save any of them.

Peter pushed the voice away.

He had to believe. Just for one more night, he had to believe.

***

Dawn came gray and cold.

The Lost Boys gathered at the fire step, checking their equipment, tightening straps, making the small adjustments that soldiers make when they're trying not to think about what comes next. The bombardment had stopped in the night, leaving a silence so profound it felt like deafness.

"This is it," the officer said. He was young—not much older than them—with a face that hadn't yet learned to hide its fear. "When the whistle blows, you go over. Walk, don't run. The wire's been cut. The Germans are finished. All you have to do is occupy their trenches."

Walk.

Don't run.

Like a Sunday stroll through the park.

"Any questions?"

No one spoke.

"Good. Fix bayonets. And good luck."

The click of metal. The rasp of blades. The sound of boys becoming soldiers, becoming weapons, becoming whatever the war needed them to be.

Peter looked at his Lost Boys.

Slightly, jaw tight, eyes forward, refusing to feel anything he couldn't control.

Tootles, one hand pressed against his chest, against Rose, against everything he was fighting for.

The Twins, shoulders touching, breathing in unison.

Curly, massive and silent, a wall between the younger boys and the void.

Keeper, too young, too small, his rifle almost bigger than he was.

Nibs, notebook tucked away, ready to remember whatever happened next.

"Whatever happens," Peter said quietly, "we stay together. We don't leave anyone behind."

"Still here."

"Still together," they answered in unison.

The whistle blew.

***

They went over the top at 7:30 a.m. Peter climbed the ladder and stepped into hell.

The wire wasn't cut. He saw that immediately—the great tangles of barbed metal still stretched across No Man's Land, intact, waiting. The bombardment hadn't destroyed it. Hadn't destroyed anything. The shells had fallen for seven days, and the wire was still there, and the German trenches were still there, and the machine guns—

The machine guns were still there.

The first burst took down the boy next to Peter. One moment walking, the next crumpled, folded, gone. The sound came after—that terrible hammering rhythm, *tick-tick-tick-tick*, the crocodile's heartbeat amplified a thousandfold.

"DOWN!" someone screamed.

But you couldn't go down. The orders were to walk. The orders were to keep moving. The orders—

The second burst took five more. Peter saw them fall—saw the way bodies move when life leaves them, the puppet-collapse, the ragdoll sprawl—and kept walking.

Because that was the order.

Because there was nothing else to do.

The field was chaos. Men were screaming, falling, crawling through mud that was already turning red. Shells were landing now—German artillery, finding their range, correcting for the masses of soldiers pouring across the open ground. The air was full of metal and fire and the whine of things designed to kill.

"PETER!"

He turned. Tootles was behind him, face white with terror, but still moving, still walking, still alive. Rose's photograph was clutched in his hand—he'd pulled it out, was holding it in front of him like a shield, like a talisman, like the only thing standing between him and the void.

"STAY CLOSE!" Peter shouted.

They pushed forward. The wire was ahead, a wall of barbed metal, uncut, unbroken, exactly where it had been for two years. Men were piling up against it, looking for gaps, finding none, becoming targets.

"HERE!" Slightly had found something—a shell crater, a depression in the earth, the smallest amount of cover in an ocean of death. "GET DOWN!"

The Lost Boys dove into the crater. Pressed themselves against the mud. Listened to the machine guns rake the ground above them, heard the bullets whisper past, felt the ground shudder with every shell.

"Is everyone—" Peter counted heads. Slightly. Tootles. The Twins. Curly. Keeper. Nibs. All there. All alive. "Thank God. We need to move. Find a gap in the wire—"

"There is no gap." Slightly's voice was flat. "The wire wasn't cut. The bombardment didn't work. This whole thing—" He gestured at the chaos around them. "This whole thing is a slaughter."

"We have to try—"

"Try what? Walk into machine gun fire? Throw ourselves on the wire?" Slightly grabbed Peter's arm. "Look around, Pan. Look at what's happening. This isn't a battle. This is murder."

Peter looked.

The field was covered with bodies. Hundreds of them. Thousands. Boys in khaki, lying where they'd fallen, some still moving, most not. The machine guns hadn't stopped. They kept firing, kept finding targets, kept ticking away the seconds of lives that had barely begun.

Tick. Tick. Tick.

The crocodile was feasting.

"We stay here," Slightly said. "We wait for dark. We crawl back. That's the only way any of us survive this."

"We can't just—"

"We can. We have to." Slightly's eyes were fierce. "I promised my mum I'd come home, Pan. And I'm not breaking that promise for some officer's fantasy about walking across No Man's Land."

Peter wanted to argue. Wanted to fight. Wanted to believe that courage and determination could overcome machine guns and uncut wire.

But he looked at his Lost Boys—at their faces, their terror, their desperate hope—and he knew Slightly was right.

"We wait," he said. "We wait for dark."

***

The shells found them at noon.

They'd been in the crater for hours, pressed against the mud, listening to the slaughter continue around them, counting the minutes until darkness might offer escape. The machine guns had slowed but not stopped. The cries of the wounded filled the air, a constant chorus of agony that became, after a while, just another kind of silence.

Tootles had Rose's photograph out. He was talking to it—quietly, steadily, a stream of words that Peter couldn't quite hear but understood anyway. Plans. Promises. The future he was building word by word, making it real through repetition.

"The garden," Tootles murmured. "The roses. The window in the kitchen where you'll stand in the mornings, watching the children play. I can see it, Rose. I can see all of it. We're so close now. Just have to make it through today. Just have to—"

The whistle came first.

Peter knew that sound. Had learned it in a thousand bombardments, had trained his body to react before his mind caught up. He was moving before he knew why—reaching for Tootles, grabbing his arm, trying to pull him deeper into the crater.

"DOWN—"

The world exploded.

Light and sound and pressure, the universe folding in on itself. Peter felt himself lifted, thrown, dropped. His ears rang. His lungs burned. He couldn't tell which way was up.

Then silence. The terrible silence that followed every shell.

Peter opened his eyes.

He was lying on his back at the bottom of the crater. Mud everywhere—on his face, in his mouth, caked in his eyes. He wiped it away. Tried to move. Everything hurt, but everything worked.

He sat up.

The crater was larger now. The shell had widened it, deepened it, reshaped it into something new. Bodies were scattered around the edges—men who had been seeking cover, who had found death instead. Some were still. Some were moving. Some were making sounds that would haunt Peter for the rest of his life.

"Slightly? SLIGHTLY!"

"Here." The voice came from his left. Slightly was pulling himself out of the mud, bleeding from a cut on his forehead but alive. "I'm here."

"Curly?"

A grunt. Curly emerged from beneath a pile of debris, hauling Keeper with him. The younger boy was crying—silent tears cutting tracks through the mud on his face—but he was alive.

"Nibs?"

"Here." Nibs was clutching his notebook. It was soaked with mud and blood, but he was holding it like it was the only thing keeping him tethered to the earth.

"Tootles?" Peter looked around frantically. "TOOTLES!"

He found him at the edge of the crater.

Billy Hawkins—Tootles—was lying on his back, his eyes open, his hand pressed against his chest. Against Rose. Against the photograph that had traveled with him through a year of war, that had been kissed and patted and talked to every night, that held all his hopes and dreams and the future he'd been building word by word.

The photograph was red now. Soaked through. And beneath it—

"No." Peter crawled to him. "No, no, no—"

"Pan." Tootles's voice was a whisper. Wet. Wrong. "Pan, is that... is that you?"

"I'm here. I'm here, Billy." Peter pressed his hands against the wound, trying to stop the bleeding, knowing it was too late, knowing there was nothing he could do. "Hold on. We'll get you to a medic. We'll—"

"Tell Rose."

"What?"

"Tell Rose—" Tootles coughed. There was blood on his lips. "Tell her I heard it. Her laugh. I heard it, Pan. Just now. Just before the shell hit. She was laughing, and it was the most beautiful—"

He stopped.

His eyes were still open, still looking at Peter, but there was nothing behind them now. Nothing but the sky reflected back.

Peter sat there, holding Billy Hawkins's body, and the world kept exploding around him, and the machine guns kept ticking, and somewhere in England a girl named Rose was laughing and didn't know yet that her whole future had just ended in a crater in France.

***

They stayed until dark.

Peter didn't let go of Billy's hand. Couldn't. Even when Slightly tried to pull him away, even when Curly physically lifted him, he kept hold of that hand—the hand that had patted Rose's photograph twice every night, the hand that had dealt cards and written letters and dreamed of holding children that would never be born.

"We have to go," Slightly said. His voice was gentle, broken. "Pan. We have to go, or we'll die too."

"I can't leave him."

"He's already gone. He's not in there anymore." Slightly knelt beside him, pressed their foreheads together. "I know. I know. But the Billy who mattered—the one who loved Rose, who made us laugh, who believed in after—he's with you now. You're carrying him. You'll always carry him."

They took his identification tags. His letters. Rose's photograph, muddied and bloodied but still showing her face, still holding her smile.

They couldn't take his body. Couldn't carry him back across that wasteland of wire and shells and the machine guns that were still, even now, still finding targets in the dark.

They left Billy Hawkins in a crater on the Somme, on the first day of the battle, along with 19,000 other British boys who would never go home.

***

The ledger entry came later.

Peter wrote it by candlelight, in a dugout that felt impossibly empty without Billy's laugh, without the nightly ritual of photograph and kiss and two pats over his heart.

Billy Hawkins. Tootles.

Age nineteen. London. Gap-toothed smile. Relentlessly optimistic. Made us believe in "after," even when after seemed impossible.

He had a girl named Rose. She laughed too loud, he said. He was going to marry her. Three bedrooms upstairs. Kitchen with a big window. Garden out back for her flowers. Children named William and Margaret and all the Lost Boys' names, one by one.

He died July 1, 1916. First day of the Somme. Shell. Not quick enough.

His last words: "Tell Rose I heard her laugh."

I have her photograph now. It's covered in his blood. I don't know if I should clean it or leave it. I don't know what she'd want. I don't know how I'm going to tell her.

I don't know how I'm going to tell anyone.

Even when we're not together. Even when the crocodile takes us one by one.

Peter closed the ledger. Pressed it against his chest.

And in the darkness of a dugout in France, surrounded by the boys he hadn't lost yet, he wept for the one he had.

Chapter Fifteen

The Keeper

France, July–November 1916

The Somme didn't end.

That was what Peter couldn't understand, couldn't accept, couldn't wrap his mind around. The first day had been a slaughter—19,000 dead before noon—and yet the battle continued. Day after day. Week after week. The same ground, the same wire, the same machine guns ticking away lives like a clock that would never run down.

July became August. August became September. The craters multiplied, overlapped, merged into a landscape that looked like the surface of a dead moon. Villages that had stood for centuries became names on maps and nothing more. The Somme ate everything—men, materiel, hope—and still the generals sent more.

The Lost Boys survived.

Not intact. Never intact. Billy's absence was a wound that wouldn't heal, a silence where his laugh should have been. They didn't talk about him—couldn't talk about him—but they felt him. Every evening when no one pulled out Rose's photograph. Every night when no one said "after" like it was a prayer. Every morning when they counted heads and came up one short.

But they survived.

Peter ran messages. Slightly kept them together with his sharp words and sharper love. The Twins held each other through every barrage, always. Curly stood watch. Nibs wrote.

And Geoffrey—Keeper—grew up.

***

It happened slowly, then all at once.

The boy who had been too young, too eager, too desperate to belong changed in the months after Billy's death. The desperate energy faded. The need to prove himself quieted. In its place came something steadier. Something stronger.

He started helping with the ledger.

"Show me," he said one evening, appearing at Peter's elbow while he was writing. "Show me how you do it."

"Do what?"

"The names. The stories. The—" Geoffrey gestured at the journal. "The keeping. I want to learn."

Peter looked at him. Really looked, for the first time in weeks. Geoffrey was sixteen now—his birthday had passed somewhere in the chaos of the Somme, uncelebrated, unacknowledged—and his face had changed. The childish roundness was gone, replaced by angles and shadows that shouldn't have belonged to someone so young.

"Why?" Peter asked.

"Because someone has to. You can't do it alone." Geoffrey's voice was steady. "And because—if something happens to you—someone needs to know how. Someone needs to be able to continue."

Peter thought of Barrie. Of the journal, passed down, given as a gift. Of the work that was bigger than any one person.

"Sit down," he said. "I'll show you."

***

Geoffrey learned quickly. He had a gift for it, Peter realized—a way of seeing people, of noticing the small details that made them real. He started carrying his own notebook, smaller than Peter's but filling up just as fast. He wrote about the soldiers they met, the boys in other units, the stretcher-bearers and medics and all the invisible workers who kept the war machine grinding.

Private Harrison. Seventeen. Liverpool. Afraid of chipmunks but not of shells. Says fear is strange like that—you can face death every day and still scream when something small crawls by you. Wants to be a teacher after. If there is an after.

Corporal Williams. Twenty-two. Cardiff. Sings in his sleep—actual songs, with words and melodies. The others used to complain but now they find it comforting. He doesn't know he does it. No one's told him.

"You're good at this," Peter told him.

"I learned from you."

"No. This is yours. Your voice. Your way of seeing." Peter touched the notebook. "When I write, I'm trying to capture what was. When you write, you're capturing what *is.* The moment. The feeling. It's different. It's valuable."

Geoffrey ducked his head, embarrassed but pleased. "I just want to help. I want to make sure no one disappears."

"That's exactly what I want too." Peter met his eyes. "You're a Keeper, Geoffrey. In every sense. Don't ever forget that."

***

The wound came in September.

Peter was running a message through a barrage—routine, by then, as routine as anything could be in hell—when a shell landed close enough to throw him. He woke in a field hospital with his left arm wrapped in bandages and a doctor frowning down at him.

"You'll live," the doctor said. "But you're done running for a while. Shrapnel damage. Needs time to heal."

"How long?"

"Weeks. Maybe months." The doctor moved to the next bed. "Consider yourself lucky. Most runners don't get second chances."

Peter lay in the hospital and thought about the ledger. About the names still being added every day, every hour. About the work that couldn't stop just because he was wounded.

Geoffrey appeared on the third day.

"Slightly sent me," he said. He was carrying Peter's journal—the original, the leather one from Barrie. "Said you'd want this."

"How did you get leave?"

"I didn't." Geoffrey grinned, a flash of the eager boy he'd been before the Somme. "I'm supposed to be on a supply run. Took a detour."

"You could be court-martialed."

"For bringing a journal to a wounded friend?" Geoffrey sat down on the edge of the bed. "I'll take my chances."

Peter took the journal. Held it against his chest. Felt the weight of all those names, all those stories, pressing against his heart.

"I can't write," he said. "My arm—"

"I know. That's why I'm here." Geoffrey pulled out his own notebook. "I'll keep the ledger while you're gone. Both of them. Yours and mine. I'll write down everyone, Pan. I won't let anyone disappear."

Peter looked at this boy—this child, really, who had grown up in the space of months, who had learned to hold death in his hands without flinching—and felt something break and mend at the same time.

"Write everything," he said. "Even the things that don't seem important. Especially those."

"I will."

"And Geoffrey?"

"Yes?"

"Come back. When you're done. Come back and tell me every name."

Geoffrey pressed his forehead to Peter's. The ritual they'd all learned from The Twins, the gesture that meant *I see you, I'm here, we're together*.

He was gone before Peter could say anything else. Back to the front. Back to the war. Back to the work that never ended.

***

Peter healed slowly. The weeks in the hospital stretched into two months. His arm mended, the shrapnel scars fading to pale lines that would mark him forever. He read letters from the Lost Boys—from Slightly, sardonic and worried; from Nibs, still writing his book; from The Twins, their notes always signed together, two names like one.

And from Geoffrey.

The letters were full of names. Pages and pages of names, each one with a story attached. Geoffrey was keeping the ledger faithfully, filling it with details Peter never would have noticed—the way a soldier held his cigarette, the particular cadence of someone's snore, the exact shade of mud on a November morning.

I understand now, Geoffrey wrote in one letter. *Why you do this. It's not about remembering the dead. It's about proving they lived. Every detail I write down is proof—proof that Harrison was afraid of squirrels, proof that Williams sang in his sleep, proof that the boy in the crater loved his dog. The war wants to erase them. The ledger says no.*

Peter read the letter three times. Then he put it with the others, in the bundle he kept tied with string, the record of his Lost Boys' survival.

Geoffrey understood.

The work would continue.

***

Peter returned to the front in November.

The Somme was winding down—not because anyone had won, but because the mud had become impassable, the cold had become unbearable, the armies had exhausted themselves against each other. The great offensive that was supposed to end the war had become just another stretch of wasteland, another set of names for the history books, another failure that would be called a victory by men who had never seen the wire.

Geoffrey met him at the communication trench entrance.

"Pan!" He looked older. Of course he did. Everyone looked older now. "You're back. You're really back."

"I'm back." Peter clasped his shoulder. "How are the others?"

Geoffrey's grin was tired but real. "Still here. I kept the ledger. Both of them. Filled almost a whole notebook. I have so much to tell you—"

"Later. Let's get inside first."

They walked through the trenches together, Geoffrey talking constantly, filling Peter in on everything he'd missed. The battles. The casualties. The small moments of humanity that had survived between the horrors.

"The Twins made a friend," Geoffrey said. "German prisoner, if you can believe it. They were on guard duty, and he started talking to them, and now they play cards together when no one's watching. Slightly says it's treason. The Twins say it's diplomacy."

"That sounds like them."

"And Curly spoke. A whole paragraph. About Emma—his sister. She sent him a photograph of herself, all grown up, and he just... started talking. Telling us about her. What she was like as a baby. How she used to follow him everywhere." Geoffrey shook his head. "It was beautiful, Pan. I wrote it all down."

"Of course you did."

They reached the dugout. The familiar smell of mud and sweat and men living too close together. The sound of shells in the distance, the eternal ticking.

The Lost Boys were there. Slightly. The Twins. Curly. Nibs. All of them, looking up when Peter entered, their faces breaking into something like relief.

Slightly stood. "Pan. About bloody time."

"I came as fast as I could."

"You came as fast as the doctors let you." Slightly embraced him, briefly and fiercely. "Don't get wounded again. We need you here."

"I'll try."

"Try harder."

Peter looked around the dugout. At his brothers. At the empty spaces where Billy should have been, where Tink's wind chimes still hung and sang.

***

The shell came three days later.

They were on a work detail—of course they were, always the work details, the endless filling of sandbags, the endless shoring of walls—when the barrage started. Not unusual. The Germans liked to shell the support trenches, keep everyone on edge, remind them that nowhere was safe.

Peter heard the whistle. The scream that meant *close.*

"DOWN!"

Everyone dove. Bodies pressed into the mud, hands over heads, the universal posture of men trying to become smaller than death.

The shell hit.

It wasn't a direct hit, but it collapsed the trench wall, and sent a cascade of mud and sandbags and shattered wood down into the space where they'd been standing.

Peter fought his way free. Dug himself out of the debris. Looked around frantically.

"SLIGHTLY!"

"Here." A hand, waving. Slightly was half-buried but alive.

"TWINS!"

"Still here." Two voices, muffled.

"CURLY! NIBS!"

Grunts. Movement. They were pulling themselves out, covered in mud, bleeding from cuts, but alive.

"GEOFFREY!"

Silence.

"KEEPER! GEOFFREY, ANSWER ME!"

Peter started digging. His hands found sandbags, wood, the detritus of the collapse. He threw it aside, dug deeper, calling Geoffrey's name over and over.

Slightly was beside him. Then the Twins. Then everyone, all of them digging, searching, praying.

They found him at the bottom of the pile.

He was curled around something—protecting it, shielding it with his body even as the walls had come down. Peter saw what it was and felt his heart crack.

The ledger.

Geoffrey had been carrying both journals—Peter's and his own—when the shell hit. And when the trench collapsed, when there was no time to run, no time to save himself, he had done the only thing that mattered.

He had protected the names.

"Geoffrey." Peter pulled him free of the debris. The boy was still breathing—barely, shallowly, the sound of lungs that weren't working right. "Geoffrey, I'm here. You're going to be fine. We'll get you to a medic—"

"Pan." Geoffrey's eyes opened. They were calm. Too calm. "I kept them safe. The names. They're all here."

"I know. I know you did. Now let us—"

"I couldn't let them disappear." Geoffrey's hand found Peter's and pressed the journals into his grip. "That's the work. That's what matters. Not me. The names."

"You matter. You matter more than—"

Geoffrey's voice was a whisper now. Fading. "I'll still be here, Pan. In the pages. With all the others. You'll write my story, won't you? You'll make sure I'm remembered?"

"Geoffrey—"

"Promise me."

Peter pressed his forehead to Geoffrey's. He felt the warmth leaving him, the life slipping away.

"I promise," Peter said. "I'll write everything. I'll make sure the world knows you were here. That you were brave. That you were the best Keeper any of us could have asked for."

Geoffrey smiled. A small smile, almost peaceful.

"Tell them—" he started.

He never finished.

***

The ledger entry was the hardest Peter ever wrote.

Geoffrey. Keeper.

Age sixteen. He lied about his age to enlist. Said he was eighteen, looked about twelve. Too young for war. Too young for death. Too young for any of this.

But he understood something that most soldiers never learn: the names matter more than the person writing them. The work matters more than the worker.

When the trench collapsed, he didn't run. Didn't try to save himself. He curled around the ledgers—mine and his—and protected them with his body. His last act was keeping the names safe.

I have his notebook now. Filled with his words, his observations, his way of seeing the world. All of it recorded. All of it saved. He died November 1916. Late Somme. The battle was almost over. The mud was turning to ice. He was sixteen years old, and he died protecting stories.

That's who he was. That's what he gave.

Keeper.

Still here. In the pages. In the names. In the work that will continue after all of us are gone.

Peter closed the ledger.

He had two notebooks now. His and Geoffrey's. Two sets of names, two ways of seeing, two voices recording the war.

The Twins were holding each other. Curly was silent, but his shoulders shook. Nibs was writing—always writing—tears streaming down his face.

And Slightly was looking at Peter with eyes that held something beyond grief. "Three," he said quietly. "Tink. Tootles. Keeper. We've lost three."

"I know."

"How many more?"

Peter didn't answer. Couldn't answer. The crocodile was still circling. The clock was still ticking.

All he could do was keep writing.

All he could do was keep the window open.

All he could do was make sure that when the war finally ended—if it ever ended—someone would know they had been here.

Someone would remember their names.

Interlude III

The Window Army

London, 1916–1917

The house was full.

Seventeen soldiers now—more than the building could reasonably hold, more than Wendy had ever imagined when she'd taken in that first broken boy. They slept in shifts, shared beds, ate in rotations. The volunteers worked around the clock, and still it wasn't enough.

It was never enough.

Samuel had been the first. Then Thomas, Edward, James, Michael—not Peter's friend, but another Michael, one of the countless Michaels the war had chewed up and spat out. Then more, and more, until the house that had once been too quiet was bursting with the sounds of men relearning how to live.

They came because Peter sent them.

He wrote their names on slips of paper and gave them to the boys being shipped home. They were the wounded, the shattered, the ones who had nothing waiting for them but empty rooms and pitying looks. *Go to 14 Wimpole Street*, the papers said. *Find Wendy Panning. Tell her Peter sent you. She'll help.*

They came with trembling hands and hollow eyes. They came speaking in languages of horror that civilians couldn't understand. They came, and Wendy opened the door, and the window stayed open, and the light kept scattering.

***

Wendy sat with the letter folded in her hands for a long time. The words blurred. For the first time since the war began, she wondered not whether Peter would come home, but whether hope itself could survive what was being asked of it.

The letter about Billy arrived in August.

Peter didn't try to soften it. Couldn't, perhaps. The words were stark, brutal, true:

Tootles is gone. July 1st. The Somme. He was talking about Rose when the shell came.

I have her photograph now. It's covered in his blood. I don't know if I should send it to her or keep it. I don't know what she'd want.

Please—if you can—find her. Tell her he heard her laugh. Those were his last words. Tell her he heard her laugh, and he died holding her picture, and he never stopped believing in "after."

I don't know how to do this. I don't know how to keep writing names when they're my brothers. I don't know how to—

The letter stopped there. Unfinished. Sent anyway.

Wendy sat in the nursery with the letter in her lap and wept.

Not just for Billy—though she wept for him too, for the gap-toothed boy who had believed so fiercely in love and futures and all the things the war was stealing. She wept for Peter. For what this was doing to him. For the son who had come back from one darkness only to walk into another.

The prism caught the afternoon light. Rainbows scattered across the walls.

Light doesn't disappear, she had told him once. *It just finds new places to land.*

But what about the boys? Where did they land when the shells took them? What window led to wherever they went?

She didn't know. She would never know.

All she could do was keep this window open. Keep this light burning. Keep catching the boys Peter sent her, the ones who had made it home with their bodies if not their souls.

***

Finding Rose was easier than Wendy had expected. The girl lived in a village outside London—a place of hedgerows and church bells and the particular quiet of communities that had given all their young men to the war. She was working as a nurse's aide at a local hospital when Wendy arrived, changing bandages on a boy who couldn't have been more than sixteen.

"Rose Whitmore?"

The girl looked up. She was pretty, dark-haired, with a smile that came easily even now, even surrounded by suffering. Wendy could see what Billy had loved.

"Yes?"

"My name is Wendy Panning. My son... my son knew Billy Hawkins."

Rose's face changed. The smile faded. The light behind her eyes flickered and steadied and became something else—something braced for impact.

"Knew," she repeated.

"Yes."

"Past tense."

"Yes."

Rose set down the bandages. Very carefully. Very deliberately. Then she walked past Wendy, through the ward, out a door that led to a small garden. Wendy followed.

They stood among roses—actual roses, climbing a trellis, blooming despite everything—while Rose pressed her hands to her face and breathed.

"He's dead," Rose said. It wasn't a question.

"Yes. July 1st. The Somme."

"I knew." Rose's voice was strange. Distant. "I felt it. That morning. I woke up and something was different. Something was missing. I told myself I was being foolish, that it was just anxiety, that—" She stopped. Breathed. "But I knew. I knew he was gone."

Wendy waited.

"How?" Rose asked.

"A shell. He was with my son. With Peter. He was—" Wendy's voice cracked. "He was talking about you. About the wedding. The house. The garden. He was holding your photograph."

"The one he took before he left. The one he kept over his heart."

"Yes."

Rose was quiet for a long moment. The roses swayed in the breeze, red and pink and white, impossibly alive in the midst of all this death.

"What were his last words?"

"He said..." Wendy had to stop. Gather herself. "He said to tell you he heard your laugh. At the end. He heard you laughing."

Rose made a sound. Not quite a sob. Something deeper, rawer, more animal.

"I was," she whispered. "That morning. Before I woke up and felt him gone. I was dreaming about him, and we were dancing, and I was laughing." She turned to look at Wendy with eyes full of something too large to name. "He heard me. Across all that distance. All that death. He heard me laugh."

"He loved you very much."

"I know. I loved him too. I still do." Rose wiped her face. Straightened. Put herself back together with the efficiency of women who had learned that grief was not a luxury they could afford. "What happened to the photograph? The one he was holding?"

"Peter has it. He didn't know what to do with it. Whether to send it or keep it."

"Tell him to keep it." Rose's voice was steady now. "I have other photographs. But Peter—Peter was with him at the end. If the picture gives him comfort, he should have it."

"I'll tell him."

"And tell him..." Rose hesitated. "Tell him thank you. For being with Billy when it happened. For not letting him die alone."

"He would never have left him. Billy was his brother."

"I know. Billy wrote about the Lost Boys. All the time. He said they were the family he'd chosen, the brothers he'd never had." Rose almost smiled. "He said Peter was going to lead them all home."

Wendy felt tears sliding down her face. "I hope he's right," she said. "I hope Peter knows the way. I hope he comes home. I hope they all do."

Rose's voice was gentle. "Some of them won't. We both know that. But Billy believed. Right up to the end, he believed. And maybe..." She looked at the roses climbing their trellis, reaching for the sun. "Maybe belief is its own kind of window. Maybe the boys who believe hard enough find their way home, even when their bodies can't."

"Is that what you believe?"

Rose was quiet for a long time.

"I believe Billy loved me," she said finally. "I believe I loved him. I believe that what we had was real, and true, and worth everything." She touched one of the roses—gently, reverently. "And I believe that somewhere, somehow, he's still laughing about that stupid gap in his teeth and waiting for me to catch up."

"Nothing is easy anymore."

"No. But we do it anyway." Rose smiled—a smile that came easily, even now, even in the midst of grief. "That's what Billy would have wanted. Keep going. Keep believing."

***

The soldiers kept coming.

Autumn. Winter. Spring. The war ground on, and the broken boys found their way to Wimpole Street, and Wendy kept opening the door.

She learned their stories. Their names. The small details that made them real. She wrote them in her own notebook—not a ledger like Peter's, but something softer. A record of healing. Of the slow, painstaking work of putting shattered boys back together.

Samuel is speaking again. Full sentences now, when he's calm. He told me about his mother today—how she used to make bread on Sundays, the way the house smelled, how safe he felt. It's the first time he's mentioned anything from before.

Thomas can sleep without screaming. Only managed two hours, but two hours is better than none. The new medication helps. So does the routine—meals at the same time, lights out at the same time, the predictability that the war stole from him.

Edward's hands have stopped shaking. He played the piano today—just a simple piece, nothing complicated—but his fingers remembered. The music is still in there, underneath the damage. We just have to help him find it.

And every night, Wendy sat in the nursery and thought of Peter. Her son, her lost boy, was still over there, still writing names, still fighting the crocodile with nothing but ink and hope.

Come home, she thought. *Please come home.*

***

The letter about Geoffrey arrived in December. Peter's handwriting was different now—shakier, cramped, as if his hand couldn't quite keep up with what his heart needed to say.

We lost Keeper.

He was sixteen. When the trench collapsed, he curled around the journals and shielded them with his body. His last words were about the names—making sure they were safe, making sure no one would disappear.

I don't know how to—

I can't—

There are so few of us left. Slightly. The Twins. Curly. Nibs. Me. That's all. Six, out of the boys who left London together. Six, and the war isn't over, and the crocodile keeps ticking, and I don't know how many more—

It's so dark here, and the ticking never stops, and sometimes I can't see the light at all.

Wendy wrote back that night. The words poured out of her, urgent and fierce:

You are not alone. You have never been alone. Every boy you send me, every name you write, every life you save with your ink and your hope—they're proof. Proof that the light doesn't disappear.

Come home when you can. Bring whoever you can bring. I'll take care of them. I'll take care of you. That's what mothers do.

She sealed the letter, sent it, and returned to the nursery.

The soldiers were settling in for the night with the quiet murmurs of men who had learned to find comfort in each other's presence. Somewhere downstairs, she heard singing. Thomas, maybe, or one of the newer arrivals.

Wendy sat in the rocking chair.

The window was open.

The prism was scattering light.

And somewhere, across the water, in a trench that smelled of mud and death, her son was writing names and waiting for the dawn.

Chapter Sixteen

The Mud

France/Belgium, 1917

They called it Passchendaele. Later, historians would argue about the name—whether it should be called Third Ypres, whether Passchendaele was just one objective in a larger offensive, whether names even mattered when so many boys were dying. But the soldiers knew what to call it.

They called it the mud.

Not mud like they'd known before. Not the Somme's churned earth, not the everyday muck of trench life. This was something else—something alive, hungry, patient. The rain fell for months without stopping, and the ground became soup, and the soup became quicksand, and the quicksand became a monster that swallowed men whole.

Peter had seen bad terrain before. But nothing—nothing—had prepared him for this.

***

"Don't step off the duckboards."

That was the first thing the guide told them. The only thing that mattered.

"I don't care if shells are falling. I don't care if the kaiser himself is shooting at you. You stay on the boards. You step off, you're dead. The mud will take you, and we won't be able to get you out."

Peter looked at the landscape around them—if you could call it a landscape. The shelling had destroyed the drainage system months ago, and now the battlefield was a swamp, a gray-brown sea of water and mud, punctuated by the splintered stumps of trees that had once been forests and the bodies of men that had once been soldiers.

Some of the bodies were floating. Some were half-submerged, faces visible, eyes open, looking at nothing.

Some were still moving.

"Christ," Slightly whispered. "Is that—"

Peter grabbed his arm. "Don't look. Just keep walking."

But Slightly couldn't stop looking. None of them could.

Because the men who were still moving weren't wounded. They weren't crawling toward help. They were sinking—slowly, inexorably, pulled down by the mud, their arms reaching toward the duckboards, their voices crying out for help that couldn't come.

"We have to—" Nibs started.

"No." The guide's voice was flat. Hard. "You can't save them. Try, and you'll go in too. The mud doesn't let go. Once it has you, it keeps you."

"But they're *alive*—"

"For now. Not for long." The guide turned and kept walking. "Stay on the boards. That's all you can do. Stay on the boards and pray you don't join them."

***

The Lost Boys stayed on the boards.

They walked in single file—Peter, Slightly, The Twins, Curly, Nibs. Six boys, all that remained of the group that had left London together. The war had taken Tink and Billy and Geoffrey. The war was still taking, still hungry, still ticking away the seconds until it could swallow the rest.

Peter wrote in his journal every night, but the words felt smaller now. Insufficient. How did you describe a landscape where the ground itself was an enemy? How did you capture the sound of men drowning in mud, their voices getting fainter, their hands getting slower, until finally there was just silence and another body for the earth to claim?

The mud is patient, he wrote. *It doesn't rush. It doesn't need to. It knows we have to cross it. It knows we'll slip eventually. All it has to do is wait.*

Hook is here. I feel him in every step. He doesn't speak in the dreams anymore—just watches. Just waits. Like the mud. Like the crocodile.

I think maybe Hook was always here. In every war. In every darkness. He's the thing that makes us forget we're human. The thing that turns boys into bodies and bodies into nothing at all.

***

Curly had saved Keeper's legacy. That's what Peter thought about, later, when he was trying to make sense of what happened. Curly had watched Geoffrey die protecting the ledgers. Had seen what it meant to choose the names over yourself. Had learned something in that moment—something about sacrifice, about purpose, about what mattered more than survival.

It was October. The assault on Passchendaele ridge had failed again—the fourth time, or the fifth, Peter had lost count—and they were retreating through the mud, trying to reach the relative safety of the support trenches before the German counterattack arrived.

The duckboards were slick with rain. Men slipped and caught themselves. Men slipped and didn't. The mud was everywhere, reaching for them, hungry. It felt like walking the plank of a pirate ship.

Nibs was carrying the notebooks. He had three of them now—Peter's ledger, Geoffrey's notebook, and his own—wrapped in oilcloth, clutched against his chest. He was exhausted, staggering, trying to keep his balance on the narrow boards while the shells fell around them.

Peter saw it before it happened. The shell, landing twenty yards to the left. The explosion. The shockwave that knocked Nibs sideways, off the boards, toward the mud.

"NIBS!"

Peter lunged for him, but he was too far away, too slow, couldn't reach—

Curly reached.

The big man moved faster than Peter had ever seen him move. He threw himself at Nibs, grabbed his arm, hauled him back onto the duckboard. For a moment, a perfect moment, they were safe—Nibs on the boards, Curly holding him, the notebooks rescued.

Then the mud took Curly's feet.

It happened so fast. One second, he was standing; the next, he was waist-deep, the solid ground beneath the boards giving way, the mud rushing up to claim him.

"CURLY!"

Peter ran toward him. Slightly grabbed his arm, held him back.

"The boards! Pan, you'll go in too!"

"I don't care—"

"He does!" Slightly's voice was fierce. "Look at him! Look at his face! He knows what he's doing!"

Peter looked.

Curly was sinking. Already chest-deep, his arms still above the surface, his face calm—impossibly calm, the face of a man who had made a decision and was at peace with it.

"The notebooks," Curly said. His voice was steady. "Did I—did he—"

"I've got them." Nibs was on his knees at the edge of the duckboard, tears streaming down his face. "Curly, I've got them, they're safe, but you have to—we have to—"

"Tell Emma." Curly was still sinking. Shoulder-deep now. "Tell her I'm sorry I won't make it home. Tell her—" His voice cracked for the first time. "Tell her I'll read all her letters. Wherever I am. I'll read them every day."

"Curly—"

His eyes found Peter's. "And Pan. Write a good story. About me. Make sure... make sure Emma knows I wasn't afraid."

"You're not going to die." Peter was struggling against Slightly's grip. "Curly, just hold on. We'll find something—a rope, a board—"

"There's no time." The mud was at his chin now. "But it's all right. It's all right, Pan. I chose this. I chose to save the names instead of myself." He smiled—actually smiled, that rare expression that so few people ever saw. "That's what Keeper did. That's what matters."

"Still here," Peter choked.

"Still together." The mud touched Curly's lips. "Always."

He didn't struggle. Didn't thrash. Just looked at them—at his brothers, his family, the boys he had spent two years protecting—and let the mud take him.

The last thing they saw was his hand, reaching toward them, fingers spread.

Then that, too, was gone.

***

They couldn't retrieve the body.

That was the worst part. Even after the shelling stopped, even after they made it back to the trenches, even after they organized a search party—there was nothing to find. The mud had swallowed Curly completely. Not even a trace remained.

"It happens," Marsh said. He had found Peter in the dugout, staring at nothing, the ledger open in his lap. "The mud takes them. All of them. That's what Passchendaele is—a grave that doesn't stop growing."

"I couldn't save him."

"No one could."

"He saved Nibs. He saved the notebooks. And I just... I just stood there—"

"You stayed on the boards. That's what kept you alive. That's what lets you keep writing." Marsh sat down beside him. "He knew, Peter. Curly knew what he was doing. He made a choice. The same choice Geoffrey made, the same choice I would make. The names matter more than the ones who carry them."

"I don't believe that."

"Yes, you do. If you didn't, you would have gone in after him. You would have died in the mud, and the notebooks would have been lost, and all those names would have disappeared." Marsh's voice was gentle. "He saved you too, you know. By making you watch. By forcing you to stay on the boards. He saved you so you could keep going."

Peter looked at the ledger in his lap. At the pages filled with names, stories, lives.

"I don't know how to write about him," he said. "About what he did. About—about watching him go under and not being able to—"

Marsh touched Peter's shoulder. "Write the truth. Write that he was brave. That he was quiet. That he spent two years protecting the people he loved, and at the end, he chose to protect them one more time. That's his story. That's what matters."

Peter nodded slowly.

He opened the ledger. Found the page. Began to write.

***

Curly. Real name unknown. I never learned it. Swore it really was his name.

He was big and quiet, and he stood guard. That's what I'll remember most. The way he positioned himself between the younger boys and danger. The way he watched doorways and scanned crowds and never quite relaxed, even when everyone else was laughing.

He had a sister named Emma. She wrote him letters every day. He wrote back every day. When she learned to read, it was so she could understand his words without help. When he died, he was thinking about her.

October 1917. Passchendaele. He saved Nibs and the notebooks. Threw himself at the mud to push them to safety. Made a choice that cost him everything.

"Tell Emma I wasn't afraid," he said.

He wasn't. At the end, when the mud was taking him, he smiled. He actually smiled.

I don't know what happens after we die. I don't know if Curly is somewhere reading Emma's letters, the way he said he would. But I know this: he chose to save the names. He chose to make sure the stories continued. And as long as someone reads these words—as long as someone knows that a boy named Curly protected the ones he loved until his last breath—he's still here.

Even in the mud.

Even at the end.

Peter closed the ledger.

The Twins were holding each other in the corner. Nibs was clutching the notebooks—his and Geoffrey's and Peter's—as if he would never let them go again. Slightly was staring at nothing, his sardonic mask finally cracked, grief leaking through the gaps.

Five left. Peter, Slightly, The Twins, Nibs.

Five out of the boys who had left London together. Five out of the Lost Boys who had sat in a greenhouse and promised to stay together.

The war wasn't over.

The crocodile was still ticking.

And somewhere in the darkness, Hook was watching, waiting, counting down the seconds until the mud could take the rest.

Chapter Seventeen

The Book

Belgium, November 1917

Nibs wouldn't let go of the notebooks. In the weeks after Curly's death, he carried them everywhere—all three of them, wrapped in oilcloth, strapped to his chest like armor. He slept with them. He ate with them. When he went to the latrine, he took them. The notebooks had become an obsession, a sacred duty, a weight he refused to set down.

"You're going to hurt yourself, carrying all that," Slightly said.

"I don't care."

"Nibs—"

"He died for these." Nibs's voice was fierce. "Curly died to save me and them. I'm not letting them out of my sight. Not until the war is over. Not until I can take them home and... and finish what we started."

"What we started?"

"The book." Nibs touched the notebooks through his coat. "I've been writing it for three years. Every detail. Every name. Every small thing that makes us real. And when this is over—if this is ever over—I'm going to put it all together. A real book. Something that will last. Something that will tell people what happened here."

Peter watched this exchange from across the dugout. Nibs had always been the writer, the observer, the one who stood slightly outside the moment so he could capture it. But this was different. This was desperation.

"Tell me about the book," Peter said. "What's in it?"

Something shifted in Nibs's his face—the desperation fading, replaced by something softer. Something like hope.

"Everything," he said. "The greenhouse, where we used to meet. The train station, when we left. The first day in the trenches—the smell, the mud, the look in the veterans' eyes." He started talking faster, the words tumbling out. "Billy and Rose. The photograph, the two pats. The way he believed in 'after' so hard you could almost see it, this shining future he was building word by word. Tink and his wind chimes. The music he made from nothing. Geoffrey and the ledger—how he died protecting it, how he understood what mattered more than any of us."

"And Curly?"

"Curly." His voice cracked. "Curly standing guard. Curly watching doorways. Curly writing to Emma every day, never missing, never forgetting. The way he smiled at the end—actually smiled, Pan, like he was at peace." He wiped his eyes. "That's what the book is. All of them. All of us. The Lost Boys and what we found in the darkness."

"What did we find?"

"Each other." Nibs met Peter's eyes. "That's the point, isn't it? The whole point. We found each other. In the worst place in the world, we found each other. And we held on. Through everything—the shells and the gas and the mud—we held on. That's the story. That's what I have to tell."

Peter thought about his own ledger. About the names, the thousands of names, each one a life compressed into ink. He had never thought of it as a book. Just a record. Just a promise kept.

But Nibs was right. It was all one story. The living and the dead, the hope and the horror. The boys who had flown toward thunder and the boys who had never come back.

"You'll finish it," Peter said. "When the war is over. You'll finish it, and people will read it, and they'll understand."

"Promise?"

"Always."

***

The attack came three weeks later.

They were holding a section of line near what was left of Passchendaele village—which wasn't much, just rubble and craters and the profound desolation of a place that had been fought over too many times. The German offensive hit at dawn.

Peter was running a message when it started. He was halfway back to the communication trenches when the barrage began, shells falling in patterns he'd learned to read, the earth shaking, the air filling with metal.

He made it to the trench. Found his way to the dugout. Threw himself inside—

And found Nibs alone.

"Where are the others?"

"Forward position." Nibs was clutching the notebooks. Always clutching the notebooks. "They got called up an hour ago. Some kind of emergency reinforcement. I was supposed to go too, but—"

Another shell, closer. The dugout shook. Dirt rained from the ceiling.

"We need to move. This position isn't safe," Peter said.

"Nowhere's safe."

"Some places are safer than others. Come on—"

The whistle. That particular descending scream.

Peter grabbed Nibs. Tried to push him toward the entrance. If they could get to the communication trench, they might—

The shell hit the dugout.

***

Peter woke in darkness. He couldn't see. Couldn't hear anything but a high, ringing tone where sounds should have been. He couldn't feel his left arm. Couldn't feel anything except pain and pressure and the weight of something on top of him.

He pushed. Struggled. Fought his way through debris—timber and sandbags and things he didn't want to think about—until finally, finally, he broke through to air.

The dugout was destroyed. The ceiling had collapsed, the walls had caved in, the whole structure had become a crater. Peter lay at the edge, gasping, trying to make his body work.

"Nibs?"

His voice was wrong. Muffled. The ringing in his ears was so loud he couldn't tell if he was speaking at all.

"NIBS!"

He started digging. His left arm wasn't working—dislocated, maybe, or worse—but he dug with his right, clawing at the debris, searching for his friend.

He found a hand first.

Then an arm. A shoulder. A face.

Nibs was alive.

"Pan?" His voice was barely a whisper. "Pan, is that—"

"I'm here. Hold on. I'm going to get you out—"

Nibs's hand found Peter's. Squeezed weakly. "No, Pan. You can't. I'm—I'm stuck. Something's—"

Peter looked. Saw the timber beam that had fallen across Nibs's chest. Saw the way it was pressing down, crushing, compressing.

"I can move it. I just need to—"

"You can't." Nibs's voice was calm. Too calm. "I felt it happen. Felt something break. I can't—I can't breathe right." He coughed. There was blood on his lips. "It doesn't hurt. That's strange, isn't it? It should hurt."

"Don't talk. Save your strength. I'll get help—"

"There's no time." Nibs tightened his grip on Peter's hand. "The notebooks. Pan. Where are the notebooks?"

Peter looked around frantically. Saw the oilcloth bundle, half-buried in debris, still intact.

"I see them. They're safe."

"Get them. Bring them here."

Peter scrambled to the bundle and pulled it free. He brought it back to Nibs, who clutched it against his chest with his free hand—the same way he'd been clutching it for weeks, the same way he'd clutched it in life.

"I never finished," Nibs whispered. "The book. I was going to—I had so much more to write—"

"I'll finish it."

"What?"

"I'll finish it." Peter pressed his forehead to Nibs's. "I'll take your notes. All of them. And when the war is over—when I get home—I'll put it together. The book about the Lost Boys. The story of what we found in the darkness."

"You can't. You're not... you're not a writer—"

"Then I'll find one. Someone who can do it justice. Someone who'll understand what we were, what we did, what we meant to each other." Peter's voice was breaking. "But I won't let it die with you, Nibs. I won't let the story disappear."

Nibs stared at him. His eyes were wet, but there was something in them—something like peace.

"You really promise?"

"I really promise. Even when we're gone."

"Even when we're gone." Nibs almost smiled. "That's what I was trying to write. All along. That's the whole point of the book. We don't disappear."

"You're still here. Right now. With me."

"Not for long." Nibs coughed again. More blood. "But that's okay. I understand now. I understand what Keeper knew, what Curly knew. The story matters more than the storyteller. The names matter more than the one who writes them."

"Nibs—"

"Tell them about me." His voice was fading. "Tell them I wanted to write a book. Tell them I watched and listened and tried to capture everything—every small detail, every real thing.

Tell them I died holding the notebooks, the same way I lived. Still writing. Still remembering."

"I'll tell them."

"And tell Mum..." He stopped. Breathed. "Tell her I'm sorry I won't be home for Christmas. She always made the best pudding. I was going to bring everyone. All the Lost Boys. She was going to meet..."

He stopped again.

Peter held him. Felt the breaths getting shallower. Felt the hand in his getting weaker.

"Still here," Peter whispered.

"Still... together..."

Silence.

The hand went limp.

And Peter sat in the ruins of the dugout, holding his friend's body, holding the notebooks that Nibs had died protecting, and screamed until there was nothing left.

***

The ledger entry came later. After the stretcher-bearers had taken Nibs's body. After Peter had been treated for his injuries—the dislocated shoulder, the cuts, the ringing in his ears that wouldn't stop. After he had found Slightly and The Twins and told them what happened.

Michael Davies. Nibs.

Age twenty. London. The writer. The observer. The one who stood gently outside the moment in order to capture it.

He carried a notebook everywhere. Battered and ink-stained and held together with string. He said he was writing a book—about all of us, about the Lost Boys, about what it meant to be lost and found and lost again.

"That's what lasts," he told me once. "The small stuff. The real stuff. Names. Stories. The things people say when they don't think anyone's listening."

He never finished the book. He died in a dugout in Belgium, crushed by a falling beam, holding the notebooks against his chest. His last words were about the story. About making sure it continued. About proving that we don't disappear as long as someone keeps telling.

November 1917. Passchendaele.

I have his notebooks now. Three of them—mine, Geoffrey's, his. I don't know if I can finish what he started. I'm not a writer. I'm just a boy with a ledger and a promise.

But I'll try. For Nibs. For all of them.

Even in the words.

Even in the story.

Even at the end.

Peter closed the ledger.

Four left. Peter, Slightly, The Twins.

Four out of all of them.

The war wasn't over. The crocodile was still ticking. And somewhere in the darkness, Hook was watching, counting the names, waiting for the story to end.

But Peter had three notebooks now. Three sets of names, three ways of seeing, three voices recording the war.

The story wouldn't end with them.

The story would continue.

Even when there was no one left to tell it.

Chapter Eighteen

Still Together

France, March 1918

The German offensive came without warning. For months, the war had been a stalemate—the same trenches, the same ground, the same endless exchange of shells and lives. Russia had collapsed, surrendered, freed up a million German soldiers to pour into the Western Front. Everyone knew something was coming. No one knew when.

It came on March 21st.

Peter was asleep in the dugout when the barrage started. Not the ordinary shelling they'd grown used to, not the random harassment that came and went like weather. This was something else. This was the world ending.

The earth shook. The walls screamed. The air itself seemed to compress, to squeeze, to become a solid thing that pressed against your chest and stole your breath.

"UP! EVERYONE UP!"

Peter was moving before he was fully awake, grabbing his kit, his rifle, the precious bundle of notebooks strapped against his chest. Slightly was beside him, then The Twins.

They stumbled out of the dugout into apocalypse.

The German guns had found them. Not just their trench—every trench, every position, every square foot of ground the British held. Shells were falling so thick they merged into a single, continuous explosion. Gas was drifting across No Man's Land in yellow-green clouds. Soldiers were running, screaming, dying, all of it happening at once, all of it impossible to comprehend.

"Gas masks!" Slightly was shouting. "GET YOUR MASKS ON!"

Peter fumbled with his mask. Got it sealed. Looked around for The Twins—

They were there. Together. Always together. Their masks were on, their hands clasped, their eyes finding each other through the glass eyepieces.

"We need to fall back!" Peter shouted through his mask. "The position's overrun!"

They ran.

Through communication trenches that were collapsing around them. Past soldiers who had stopped moving and soldiers who would never move again. Past the ruins of positions they'd held for years, positions that were being erased in minutes by the sheer weight of German fire.

Peter didn't know where they were going. Didn't know if there was anywhere to go. He just ran, and Slightly ran beside him, and The Twins ran behind them, and they kept running until they found a crater that wasn't full of water or bodies, and they dove in.

"What's happening?" Slightly's voice was muffled through his mask. "This isn't just an attack. This is—"

"A breakthrough." Peter had heard the officers use the word, always as a theoretical, always as something that happened to other armies in other wars. "They're breaking through."

"Breaking through to what? There's nothing behind us but more trenches!"

"And behind those trenches is France. Paris. The Channel." Peter's voice was hollow. "If they break through, they win. It's over."

Silence. The shells continued to fall.

"Then we stop them," Thompson said.

"We stop them," Thomson agreed.

"We stand here, and we fight, and we don't let them past."

***

The Germans came at noon.

The barrage had lifted—not stopped, just moved, searching for targets farther back, giving the infantry room to advance. And advance they did. Peter watched them come through the smoke and gas: gray uniforms, bayonets fixed, moving with the discipline of soldiers who had been waiting months for this moment.

"Here they come!" Slightly was up on the rim of the crater, rifle ready. "Everyone—"

The first wave hit them like a breaking wave hits a shore.

Peter fired. Reloaded. Fired again. The world narrowed to the simple mechanics of killing: aim, squeeze, chamber, repeat. He saw Germans fall. He saw British soldiers fall. He saw men die in ways that would haunt him forever, if he survived long enough to be haunted.

The Twins fought side by side. Of course they did. They had never been apart, not in all the years of war—and they weren't going to start now. Peter glimpsed them in the chaos, covering each other, moving in perfect synchronization, two halves of one whole even in the midst of death.

"Tiich!"

Peter turned.

Thomson was screaming—actually screaming, a sound Peter had never heard from either twin. He was pointing at something in the crater, something Peter couldn't see through the smoke.

Peter scrambled toward him.

Thompson was on the ground. There was red on his chest, spreading, blooming like the roses Billy used to talk about. His eyes were open, fixed on his cousin, his hand reaching.

"No." Thomson dropped beside him. "No, no, no—"

"It's all right, Koi. It's all right." Thompson's voice was faint. Wet.

"You can't leave me. We're a package deal. Both of us or neither." Thomson was pressing his hands against the wound, trying to hold the blood in, trying to hold his brother together. "You promised. We promised."

"I know." Thompson reached up. Touched his forehead with his own. One last time. "Still here."

"Still together."

"Always," Thompson smiled. A gentle, peaceful smile. "I'll wait for you. Wherever I'm going. I'll wait."

"Don't—"

"Take care of yourself. Take care of Peter. Finish the story." Thompson's eyes were fading. "I love you. I always loved you. From the very beginning."

"I love you too. From the very beginning."

The light went out of Thompson's eyes.

Thomson stayed there, forehead pressed to forehead, hands still trying to stop blood that no longer mattered. He didn't move. Didn't speak. Didn't seem to notice when Slightly called his name, when Peter tried to pull him away, when the German counterattack began.

"Thomson." Peter was shaking him. "Thomson, we have to move. They're coming back. We have to—"

"Go." Thomson's voice was empty. "Go without me."

"We don't leave anyone behind. That's the rule."

"The rule was for the living." Thomson still hadn't looked up from his cousin's face. "I'm not living anymore. I'm just... waiting."

"Waiting for what?"

"For him to come back. He said he'd wait for me." Thomson finally looked at Peter, and his eyes were ancient, devastated, already gone. "Now I'll wait for him."

"Thomson—"

"Go, Pan. Save yourself. Save Slightly. Finish the story." He looked back at his brother. "I'll catch up. When it's time."

Another shell, closer. The Germans were advancing again.

Peter wanted to stay. Wanted to drag Thomson to safety by force. Wanted to do something, anything, to change what was happening.

But Slightly grabbed his arm.

"We have to go. Now."

"I can't leave him—"

"He's already gone." Slightly's voice cracked. "Look at him. He died the moment Thompson did. The body just hasn't caught up yet."

Peter looked.

Thomson had laid down beside his cousin. Had taken Thompson's hand in his. Had pressed their foreheads together one last time.

He closed his eyes.

The Germans found them like that, three hours later. Two boys, side by side, hands clasped, foreheads touching. The medics who processed the bodies said they looked peaceful. Said it was hard to tell where one ended and the other began.

They were buried in the same grave.

Of course they were.

It was the only way they'd ever wanted it.

***

Thomson and Thompson. The Twins.

Koi and Tiich. They were always just "The Twins"—two halves of one whole, inseparable since before they were born, I would imagine.

They had a ritual. A forehead touch. "Still here," one would say. "Still together," the other would answer. They did it every morning, every night, every time they came back from a mission. It was how they checked that both were real. That neither had dreamed up the other.

We all learned it from them. The rest of the Lost Boys. It became our greeting, our password, our prayer.

March 1918. German Spring Offensive. Thompson took a bullet in the chest. Died in his cousin's arms. Said he would wait.

Thomson didn't run. Didn't try to save himself. Just lay down beside his cousin and waited to catch up.

They found them together. Buried them together. I don't know if there's an afterward, a place where soldiers go when the war takes them. But if there is—if Thompson was right—they're together now. Forever.

Peter closed the ledger.

Two left. Peter and Slightly.

Two out of all of them.

The war was still raging. The German offensive was still pushing forward, eating ground, swallowing positions that had been held for years. The crocodile was still ticking.

But Peter had the notebooks. Had the names. Had the story that would continue even when there was no one left to tell it.

"Pan." Slightly's voice was soft. They were huddled in a support trench, waiting for orders, waiting for the world to stop ending. "I need to tell you something."

"What?"

"If I don't make it. If the war takes me before it takes you..." Slightly's eyes were bright with something Peter couldn't name. "Tell my mum. Tell Lilly Ann. Everything. Not just that I died—that's what the telegram will say. Tell her who I was. What we were. The rituals and the names and the stupid forehead thing. Make sure she knows I was a Lost Boy."

"You told me this before. In 1915."

"I know. But I'm telling you again. Because it matters." Slightly whispered. "I need to know you'll do it. I need to know she wo"t just get a telegram and a medal. I need to know someone will tell her the real story."

"I promise."

"All the way to the end."

And in a trench in France, with the German army pressing forward and the war grinding toward its final, bloody conclusion, two Lost Boys held on to each other and waited for whatever came next.

Chapter Nineteen

The Last Night

France, Summer–Autumn 1918

The offensive broke. Not all at once—nothing in this war happened all at once—but gradually, inexorably, the German advance ground to a halt. They had pushed too far, too fast, outrunning their supplies, exhausting their reserves. The Americans were arriving by the hundreds of thousands, fresh and eager, untouched by four years of grinding death. The balance was shifting.

By August, the Allies were advancing.

Peter barely recognized the experience. For three years, he had known only retreat and stalemate—the same trenches, the same ground, the endless back-and-forth of a war that seemed designed to last forever. Now, suddenly, they were moving forward. Taking ground. Passing through villages that had been German for years, seeing people who wept and cheered at their arrival.

"It's ending," Slightly said one evening. They were billeted in a farmhouse that still had most of its roof, a luxury that felt almost obscene after years in the mud. "Can you feel it? The war is ending."

"Don't say that."

"Why not?"

"Because I've seen too many boys die believing they were almost home." Peter was writing in his ledger, as he did every night, recording the names of the day. "The war ends when it ends. Not before."

"You're no fun anymore."

"I was never fun."

Slightly almost smiled. "That's true. But you used to believe things. Hope things. Now you just—"

"Now I just what?"

"Now you just wait. Like you're expecting the worst." Slightly looked at him with something like concern. "You still have the nightmares?"

"Every night."

"Hook?"

"Hook." Peter set down his pen. "He's closer now. In the dreams. Before, he was always in the distance, watching. Now he's right there, breathing down my neck. I can hear the ticking even when I'm awake."

"That's not Hook. That's just the war."

"Maybe they're the same thing."

Slightly was quiet for a moment. "We're going to make it, Pan. Both of us. We're going to walk out of this war and go home and I'm going to dance at Tootles's wedding—" He stopped. Corrected himself. "I'm going to visit Tootles's grave. And Rose's house. And I'm going to tell her everything, the way I should have before.

And then I'm going to go home to my mum, and she's going to hit me and hug me and force-feed me four dinners in a row."

"That sounds nice."

"It does, doesn't it?" Slightly's voice softened. "What are you going to do? When you get home?"

Peter thought about it. About the window in the nursery, the prism, the light scattering across walls he hadn't seen in three years.

"I'm going to stand at the window," he said. "Just stand there and feel the light come through. Remember what it was like to be a boy who chased rainbows."

"And then?"

"Then I'm going to find everyone's families. All the Lost Boys. I'm going to tell them what happened—not the telegram version, the real version. I'm going to make sure they know their sons were heroes. That they were loved. That they didn't disappear into the mud without anyone noticing."

"That's a lot of doors to knock on."

"I have a lot of names in my book."

Slightly nodded slowly. "And Lilly Ann? My mum?"

"First door. I promised."

"You did." Slightly reached out and clasped Peter's hand. "All the way to the end."

"All the way to the end."

***

John Darling found his wings in August. Peter knew him only through hearsay—a boy from another unit who had been attached to theirs for a few weeks during the Somme. He had talked about flying constantly: the RFC, the aircraft, the dream of seeing the war from above.

"Up there, it's different," John had said. "Up there, you can see the whole battlefield. The patterns. The beauty, even—because from high enough, even mud looks like it means something. And for a few minutes, you forget. You forget the trenches and the gas and all of it. You're just... flying."

He had finally transferred in spring of 1918, traded his infantryman's kit for a pilot's, gone up into the sky he'd been watching for years.

Peter learned of his death through the letter network that connected soldiers across the front—scraps of information passed hand to hand, names traded like currency.

"John Darling. RFC. Shot down August 12th. North of Amiens."

"Did he... was it quick?"

"They said his plane caught fire. Went down behind our lines. They found him in the wreckage." The messenger's voice was flat. "But he was smiling. That's what they said. When they pulled him out, he was smiling."

Peter wrote it in his ledger that night:

John Darling. Age twenty. Wanted to fly. Spent three years in the trenches watching the sky, waiting for his chance.

August 1918. He finally got his wings. Flew for three months. Saw the war from above, the way he'd always dreamed.

Shot down over Amiens. They found him in the wreckage, still smiling.

He flew at the end. He finally flew.

***

The armistice rumors started in October.

At first, no one believed them. The war had been going on forever; it would go on forever. That was just how the world worked. But the rumors persisted of German envoys, peace negotiations, an end that might actually be coming.

“Eleven-eleven-eleven,” someone said in the mess. “That's what I heard. November 11th, 11 a.m. The fighting stops.”

“Bollocks.”

“It's true. The kaiser's abdicated. The Germans are asking for terms.”

“They asked for terms in 1916. Nothing came of it.”

“This is different. They're beaten. Really beaten. Look around—we've been advancing for months. They've got nothing left."

Peter listened to the rumors and felt nothing. Hope had become a language he'd forgotten how to speak. All he could do was write names, keep the ledger, wait for whatever came next.

November arrived. The offensive continued. The rumors grew louder.

***

And on November 10th, a boy named Tomkins got word that his unit was being sent forward one last time.

Peter met Tomkins in the support trench, an hour before the assault. He was young—impossibly young, the kind of young that made Peter's heart ache. Sixteen, maybe, lying about his age the way Peter had lied, the way Geoffrey had lied, the way too many boys had lied because the recruiters didn't want to look too closely.

"I'm going home tomorrow," Tomkins said. He was grinning—gap-toothed, like Billy, with the same relentless optimism that hurt to look at. "War's almost over. One more day. Then I'm going home."

"Where's home?"

"Sheffield. My mum's a baker. Makes the best bread you ever tasted." Tomkins patted his pocket. "She sent me a letter last week. Said she's going to make a cake when I get back. Chocolate. My favorite."

"Sounds nice."

"It will be. It'll be brilliant." Tomkins's eyes were bright. "I can't wait, you know? To see her again. To eat something that isn't bully beef. To sleep in a real bed." He laughed. "God, remember real beds? I've almost forgotten what they feel like."

"I remember."

"Of course you do. You've been out here how long?"

"Three years. Almost four."

Tomkins whistled. "Christ. How do you do it? How do you keep going?"

Peter thought about it. About the ledger, the names, the promise he'd made to all the boys who'd died. About the window, still open, still waiting. About Wendy, still there, still believing.

"You find something to hold on to," he said. "Something that matters more than the war. More than the fear. You hold on to it, and you don't let go."

"What do you hold on to?"

"Names. Stories. The people I loved who aren't here anymore." Peter touched the notebooks strapped to his chest. "I keep them with me. Carry them everywhere. As long as I'm carrying them, they're still here."

Tomkins nodded slowly. "I carry my mum's letter. Same idea, I suppose."

"Same idea."

"Still here," Tomkins said, testing the words. "Is that what you say? I've heard people in your unit say it."

"Still here. Still together. It's—it's how we remind each other we're real."

"I like that." Tomkins grinned again. "I'm going to tell my mum about that. When I get home."

"You do that."

The whistle blew. Time to move forward.

"Good luck," Peter said.

"You too." Tomkins clasped his hand. "See you on the other side."

Peter watched him go. Watched him climb the ladder, disappear over the top, and join the wave of soldiers advancing toward German lines that might or might not still be defended, in a war that might or might not be almost over.

He didn't see the shell.

But he heard it. That whistle. That descending scream.

And when he reached the forward position an hour later, carrying messages that suddenly didn't matter, he found what was left of Tomkins in a crater twenty yards from the German wire.

The letter from his mother was still in his pocket. Still legible, somehow, despite the mud and the blood.

My darling Robert, I can't wait to see you. The cake is already baking in my mind. Come home safe. Come home soon. All my love, Mum.

Peter took the letter. Added it to the notebooks.

And that night, in a dugout that smelled like death and endings, he wrote:

Robert Tomkins. Sixteen. Sheffield. Lied about his age. After the first year, the ages stopped surprising him. Good at cards.

His mother baked bread. Made the best in Sheffield, he said. She was going to make him a chocolate cake when he got home.

November 10, 1918. The last night of the war. He died believing he was going home tomorrow.

He was. Just not the way he thought.

Peter closed the ledger.

One day left.

One more day, and the war would be over.

One more day, and the dying would stop.

He didn't believe it. Couldn't believe it. But somewhere, deep in his chest, a small, terrified part of him dared to hope.

Tomorrow, he thought. *Tomorrow we go home.*

He fell asleep holding the notebooks against his chest.

And in his dreams, Hook was there, closer than ever, counting down the final hours until the clock struck eleven.

Chapter Twenty

Armistice

France, November 11, 1918

The war would end in six hours and thirteen minutes.

Peter Panning did not know this yet. He knew only that he was cold, that his boots had not been dry in eleven days, and that somewhere in the darkness beyond the trench wall, German machine guns waited with the patience of wolves.

He knew that his hands were shaking. They had been shaking for months now. It was a fine tremor he could not control, could not hide, could not stop. Slightly pretended not to notice. That was the kindness of the trenches: you learned not to see the things that would break you if you looked too closely.

The ledger was open in his lap.

He had filled three of them now—three leatherbound books dense with names, with stories, with the small details that made a boy a person instead of a casualty report.

Thousands of names. Thousands of boys who had flown toward thunder and never come back.

Peter traced the most recent entry with his finger. The ink was barely dry.

Andrews, J. Sixteen. Cornwall. Had a great belly laugh. Died November 10, 1918. Shell. Quick.

Quick. As if that mattered. As if the speed of dying made the death smaller, easier, more bearable.

Nothing made it bearable.

***

"Can't sleep?"

Slightly's voice, hoarse with exhaustion, came from the darkness beside him.

"Can you?"

"No." A pause. "Heard a rumor."

"There's always a rumor."

"This one's different." Slightly shifted, and Peter heard the wet squelch of mud. "Runner came through an hour ago. Said the Armistice is signed."

Peter went very still.

"Signed?"

"At five o'clock this morning. In a railway carriage, somewhere in France. The war ends at eleven." Slightly's voice was strange, caught between hope and disbelief. "Six hours, Pan. Six more hours, and it's over."

Six hours.

Peter looked at his hands. At the tremor he couldn't stop. At the ledger in his lap, full of names that would never go home.

"I don't believe it."

"Neither do I. But the runner seemed certain." Slightly moved closer, pressing against Peter's side. "What if it's true? What if we actually make it?"

"Don't."

"Don't what?"

"Don't hope. Not yet. Not until we hear the guns stop." Peter closed his eyes. "I've seen too many boys die believing they were almost safe."

"I know. I know, Pan. But—I" Slightly's voice cracked. "Six hours. We've survived three years. Surely we can survive six more hours."

Peter didn't answer.

He was thinking of Tomkins. Of the chocolate cake his mother would never bake. Of all the boys who had died on the last night, the last week, the last month of a war that was already ending.

The crocodile didn't care about armistices.

The clock kept ticking until it didn't.

***

The order came at seven.

Peter was half-asleep, leaning against Slightly for warmth, when the lieutenant appeared at the entrance to their position.

"Advance on the German positions. Secure the ridge before the ceasefire takes effect."

Peter stared at him.

"Sir—the war is over. It ends at eleven."

"It ends at eleven. It's seven now. That gives us four hours to take objectives that will strengthen our negotiating position."

"Negotiating position? The war is *over*. What's there to negotiate?" Slightly's voice was raw.

"That's above your pay grade, Private." The lieutenant's voice was cold. "Orders are orders. We advance at 0730."

He walked away before anyone could respond.

Slightly was shaking. Not with cold—with rage.

"He's mad," he whispered. "They're all mad. We're going to die for—for *nothing*—"

"We're not going to die." Peter's voice was calm. Distant. The voice of someone who had seen too much to feel anymore. "We're going to follow orders. And we're going to survive. Three and a half more hours. That's all. Three and a half more hours, and then it's over."

"And if we don't? Survive?"

Peter pulled out his ledger.

"Then I'll write your name," he said. "And the world will know that you died on the last day. That you were stolen by a war that had already ended. That Hook took you—not because he had to, but because he wanted one more victory. One more moment of glory before the curtain fell."

"Hook?"

"The generals. The politicians. The men who send boys to die so they can add another medal to their chest, another line to their history books." Peter's jaw was tight. "They're Hook. They've always been Hook. And right now, Hook wants to be the hero of his own story. He wants to end the war with a victory, even if it costs—"

He stopped. Couldn't finish.

Even if it costs everything.

***

They went over the top at 0730.

It was, Peter thought, the most pointless attack of the war. The Germans barely resisted—they knew the Armistice was coming, knew there was no point in dying for a few more hours of stalemate. Most of them simply surrendered, hands raised, eyes full of the same exhausted disbelief that Peter felt.

They took the ridge by nine.

"Dig in," the lieutenant ordered. "Hold the position until the ceasefire."

Two hours. Two more hours.

Peter found a spot in the captured trench and sat down. His legs were shaking. His whole body was shaking. The adrenaline was fading, and in its place came something vast and terrifying: hope.

We made it.

We actually made it.

Slightly dropped down beside him.

"This is it, isn't it? We're actually going to survive."

"Don't say it."

"I have to say it. I have to hear it out loud." Slightly grabbed Peter's hand. "Pan. We're going to go home. Both of us. We're going to walk out of this war and see our mothers and sleep in real beds and never—*never*—have to hear another shell as long as we live. This is the war to end *all* wars."

Peter looked at his friend. At Slightly—Charlie Morton—the sardonic boy from London who had become his shadow, his brother, his partner through three years of hell.

"What are you going to do first?" Peter asked. "When you get home?"

Slightly's face softened.

"Hug my mum. Pick her up off her feet—she hates that—and just hold her until she hits me." He almost smiled. "She's going to cry and call me an idiot and probably make me eat three dinners in a row."

"She sounds—"

The shell came without warning.

There was no whistle, no scream, no time to react—just a sound like the world tearing apart and then Peter was on his back, ears ringing, dirt raining down, something wet on his face that might have been mud and might have been—

"CHARLIE!"

Peter scrambled through the debris. His hands found fabric, then flesh, then—

Slightly was lying on his side, his eyes open, his chest—

His chest was wrong. Everything was wrong.

"No. No, no, no." Peter pulled him up, cradled him, pressed his hands against the wound that couldn't be stopped. "Charlie. Charlie, stay with me."

Slightly's eyes found his. They were calm. Too calm.

"Pan." His voice was a whisper. Wet. "What time—"

Peter looked at his watch. The face was cracked, but the hands still moved.

"Ten fifty," he said. "Ten minutes. Just ten more minutes, Charlie, please—"

Slightly's hand came up, trembling, and touched Peter's forehead. The way The Twins used to do.

"Still here?" Slightly whispered.

Peter pressed his forehead to Slightly's. Felt the warmth leaving him. Felt the life slipping away like water through his fingers.

"Still together," he said. "Always. I swear."

Slightly smiled. A real smile. The one Peter had seen a thousand times in dugouts and trenches and moments of desperate hope. The smile of a boy who had been loved.

"Tell my mum—"

"I will. I'll tell her. Everything. I promise."

"Tell her I—"

He didn't finish.

His hand dropped.

His eyes stayed open, still looking at Peter, still calm, still almost smiling.

And Peter Panning held his brother in a trench in France and screamed until his throat tore, and the shells kept falling, and somewhere in the distance a clock was ticking down the final minutes of a war that had already taken everything, everything, everything—

***

At eleven o'clock, the guns stopped.

Peter didn't notice at first. He was still holding Slightly—still holding Charlie Morton, the boy from London, the boy whose mother was named Lilly Ann, the boy who had promised to dance at Billy's wedding and never would.

He didn't notice the silence spreading across the battlefield like ripples in a pond. Didn't notice the soldiers around him lowering their weapons, looking at each other in disbelief, some of them weeping, some of them laughing, all of them struggling to understand that it was finally, finally over.

He only noticed when someone touched his shoulder.

"Private. Private Panning."

Peter looked up. An officer was standing over him, his face gray with exhaustion and something that might have been sympathy.

"The ceasefire has begun. The war is over."

"The war is over," Peter repeated. The words didn't make sense. None of it made sense.

"Your friend—"

"His name was Charlie. Charlie Morton. He died at ten fifty. Ten minutes before—" Peter's voice broke. "Ten minutes. He almost made it. He almost—"

The officer was silent for a long moment. "I'm sorry," he said finally. It was the most inadequate thing anyone had ever said.

Peter looked down at Slightly's face. Still calm. Still almost smiling.

He closed Slightly's eyes. Pressed their foreheads together one last time.

And then, very carefully, very gently, he laid his brother down in the mud of a French battlefield, and stood up, and walked away from the war.

***

The ledger entry was the last one Peter would write in France.

He wrote it that night, in a ruined farmhouse, surrounded by soldiers who were celebrating and weeping and trying to remember what it felt like to be alive without being afraid.

Charles "Slightly" Morton.

Twenty-One. London. Sardonic, brilliant, the best friend I ever had.

He survived four years. Survived the Somme. Survived Passchendaele. Survived the final advance.

Died November 11, 1918. 10:50 a.m.

Ten minutes before the ceasefire.

His mother's name is Lilly Ann Morton. I have to tell her. I promised.

His last words were "Tell her I—"

He didn't finish. I don't know what he wanted to say. I'll never know. But I think it was "I love her." I think that's what he was trying to say.

Still here. Still together.

Always.

Peter closed the ledger.

He was the only one left.

The only Lost Boy who had made it through. The only one who would walk back through the window, would stand in the nursery where the prism scattered light, would hold his mother and feel the weight of everyone he'd lost pressing against his chest.

The war was over.

The dying had stopped.

But the carrying—the carrying would never stop. Not for Peter. Not as long as he lived.

He had promised to remember them.

He had promised to tell their stories.

He had promised to keep the window open, for all the lost boys who would never find their way home.

And Peter Panning, alone in a world that was suddenly, impossibly, at peace, pressed the ledger against his heart and began the long journey back to the light.

Interlude IV

The Bells

London, November 11, 1918

The bells began at eleven. Wendy was in the nursery when she heard them, standing at the window, as she did every morning, checking that the prism was catching light, that the rainbows were scattering, and that everything was ready for Peter to find his way home.

At first, she didn't understand. The bells of London had been silent for four years, muffled by wartime regulations, stilled by the same grief that had stilled everything else. She had almost forgotten what they sounded like—that cascade of sound, that tumbling joy, that ancient language of celebration.

Then the shouting started.

"IT'S OVER! THE WAR IS OVER!"

Wendy moved to the window and looked down at the street below. People were pouring out of houses, out of shops, out of every door and window. They were embracing strangers, weeping, laughing, and dancing in the streets. Someone had found a flag—a

Union Jack, faded and wrinkled—and was waving it overhead while others cheered.

The war was over.

She should have felt joy. Should have felt relief, elation, and the kind of overwhelming gratitude that was spilling across the faces of everyone below.

Instead, she felt only fear.

Where is Peter? Is he alive? Did he survive the last battle, the last shell, the last moment before the guns stopped?

She wouldn't know for days. Maybe weeks. The letters took time, and the chaos of armistice would only make them slower.

All she could do was wait.

All she could do was keep the window open.

***

The soldiers found her that afternoon. Not Peter—not yet—but the others. The broken boys she had been collecting for three years, the ones who lived in her house and learned to be human again. They gathered in the parlor, drawn by the bells, by the shouting, by the impossible news that the thing they had survived was finally, finally over.

Samuel was there. He had been the first, the one who couldn't remember his own name. Now he stood by the window, tears streaming down his face, whispering something Wendy couldn't hear.

Thomas was there. Edward. James. Michael, the countless boys the war had taken and returned in pieces.

They looked at her with eyes full of something she couldn't name.

"It's over," Samuel said. His voice was stronger now than it had been three years ago, but it still trembled. "Mrs. Panning. It's really over."

"Yes. It's really over." Wendy's throat was tight.

"We survived."

"You survived."

"Because of you." Samuel crossed the room and took her hands. "Because you never stopped believing we could find our way back."

"I just did what any mother would do."

"No. You did what no one else would." Samuel's grip tightened. "You saw us. When everyone else looked away—when our own families couldn't bear to see what we'd become—you saw us. You remembered we were still human."

Wendy thought of Peter. Of the boy who had been taken, who had found his way home through darkness, who had spent four years in another kind of darkness trying to do the same for others.

"I learned it from my son," she said. "I just hold the light."

Samuel released her hands. "Then keep holding it. Keep holding it until he comes home."

"I will."

"Promise?"

"I promise." Wendy looked around at the soldiers—her soldiers, her broken boys, her proof that the light never truly disappeared. "I promise I'll keep holding it. For Peter. For all of you. For everyone who's still trying to find their way back."

The bells kept ringing.

The cheering kept swelling.

And Wendy Panning stood in her house full of lost boys and waited for the one she loved most to come home.

***

The telegram arrived three days later.

Wendy saw the boy on the bicycle, saw him turn up her street, saw him stop at her door. She was moving before he knocked—flying down the stairs, throwing open the door, snatching the paper from his hands with fingers that shook too badly to hold it steady.

PRIVATE PETER PANNING STOP ALIVE STOP RETURNING ENGLAND DECEMBER STOP

Alive.

He was alive.

Wendy sank to her knees on the doorstep, the telegram clutched against her chest, and wept. She wept for Peter—for the boy who had survived, who was coming home, who would stand at the window again and watch the light scatter. She wept for the boys who hadn't survived—for Tink and Billy and Geoffrey and Curly and Michael and The Twins and all the others whose names filled Peter's letters, whose stories she had learned to carry alongside her own grief.

She wept for Slightly.

She didn't know yet that he was dead. Wouldn't know until Peter came home and told her. But somewhere, in the part of her heart that had learned to feel the shape of loss before it arrived, she knew.

Not all of them had made it.

Not all of them ever could.

***

She wrote to Lilly Ann Morton that night.

Dear Mrs. Morton,

I'm writing because I received word today that Peter is alive. He's coming home in December. And I know—I know in my heart—that when he arrives, he will have news about Charlie. News that he will want to deliver himself, in person, the way he promised.

I don't know what that news will be. I pray it's good. I pray your son is alive, that he's coming home too, and that this letter is unnecessary worry from a mother who has learned to fear the worst.

But if it isn't—if the news is what I'm afraid it might be—I want you to know that you're not alone. You have never been alone. Peter loved Charlie like a brother. He would never have let him face the darkness without someone beside him.

Whatever comes next, whatever Peter tells you when he arrives, please know: your son was loved. Your son was remembered. Your son was part of something bigger than himself—a family of lost boys who found each other in the worst place in the world and held on.

With love and hope,

Wendy Panning

She sealed the letter. Sent it. Returned to the nursery.

The window was open. The prism was catching the last light of the day, scattering rainbows across the walls.

"Come home, Peter," Wendy whispered.

The light shifted. The rainbows moved.

And somewhere, across the water, a boy who had survived the war was beginning the long journey back to the window that had never closed.

Part III

The Window Stays Open

Chapter Twenty-One

Coming Home

London, December 1918

The train pulled into Victoria Station on a Tuesday afternoon. Peter stood at the window of his compartment and watched London emerge from the winter fog—the rooftops and chimneys, the distant parks, the spires of churches that had been ringing their bells for a month now, celebrating an end that still didn't feel real.

He was twenty-one years old. He had left at seventeen. Four years—four years of mud and shells and the constant tick of the crocodile—and now he was coming home.

The notebooks were strapped against his chest. Three of them now: his ledger, Geoffrey's observations, Michael's unfinished book. He had carried them across France, through the chaos of demobilization, onto the ship that brought him back to England. He would carry them for the rest of his life.

The train stopped. The doors opened. Soldiers poured onto the platform, met by families who had been waiting, hoping, praying for this moment. Peter watched reunions unfold around him—mothers clutching sons, wives kissing husbands, children lifted into arms they barely remembered.

He scanned the crowd.

And there she was.

Wendy Panning stood at the edge of the platform, her face pale, her hands clasped in front of her, her eyes searching every face that passed. She looked older—of course she did, years older, years of worry and grief and the exhaustion of mothers who send their children to war. But she was still Wendy. Still his mother. Still the woman who had kept the window open.

Their eyes met.

Peter didn't remember crossing the platform. Didn't remember pushing through the crowd. He only remembered the moment when her arms closed around him and he was home, finally home, and the weight of everything he'd carried came crashing down.

"Peter." Her voice was broken. "My Peter. My boy."

"Mama." He was crying. Couldn't stop. Didn't want to stop. "I'm so sorry—"

"For what?"

"For leaving. For being gone so long. For..." He pulled back, looked at her face. "For not bringing them with me."

Wendy's expression shifted. She had known, of course. Had felt it in her bones the way mothers always do.

"How many?" she asked quietly.

"All of them." Peter's voice cracked. "Tink. Billy. Geoffrey. Curly. Michael. The Twins. And Charlie—" He stopped. Couldn't say it. "Slightly. He died on the last day. The last hour. Ten minutes before the ceasefire."

Wendy closed her eyes. Tears slipped down her cheeks, but she didn't make a sound.

"I held him," Peter continued. "At the end. I held him and we said the words and then he was gone. And I couldn't—I couldn't save him—"

"You were with him." Wendy opened her eyes and took his face in her hands. "You were with him, Peter. He didn't die alone. None of them died alone. That's what you gave them. That's what matters."

"It's not enough."

"No. It's never enough." She pulled him close again. "But it's what we have. It's all we have. And now—now you're home. And we carry them together."

They stood there for a long time, holding each other, while the platform emptied and the fog crept in and the bells of London rang their endless, joyful, inadequate song.

Then, Peter's father arrived. Paused when he saw them together, then embraced them both and pulled them to his chest. "My boy. You're home," he cried. "Your mother has been at the station every day for weeks waiting. When she didn't turn up, I knew you had made it."

"Papa," Peter cried.

"Finally, you're home, again."

***

The house on Wimpole Street was full of soldiers.

Peter had known, from his mother's letters, that she had been taking in the broken boys—the ones he had sent her, the ones who needed a window to find their way back. But knowing was different from seeing.

They were everywhere. In the parlor, playing cards. In the kitchen, helping with meals. In the corridors, walking with canes or wheelchairs or the careful steps of men learning to trust their bodies again. The house that had once been too quiet was bursting with life—damaged, healing, stubbornly persistent life.

"You did this," Peter said. He was standing in the doorway of the nursery, looking at his mother, at the soldiers who had become her family. "You built this."

"We built it. You and me." Wendy touched his arm. "You sent them to me. I just opened the door."

"The window."

"Yes. The window." She smiled. "It's still open, you know. I never closed it. Not once, in four years."

Peter walked to the nursery window. Looked at the prism, still hanging from its chain, still catching the light, still scattering rainbows across the walls.

He had dreamed of this moment. In the trenches, in the dugouts, in the long dark hours when the shells were falling and the crocodile ticked. He had dreamed of standing here, feeling the light on his face, being the boy who chased rainbows instead of the man who buried them.

But he wasn't that boy anymore.

He would never be that boy again.

"I have to go see Lilly Ann," he said.

Wendy nodded. She had been expecting this.

"When?"

"Tomorrow. First thing." Peter touched the notebooks strapped to his chest. "I promised Charlie. I promised I would tell her everything."

"Do you want me to come with you?"

"No. This is... is something I have to do alone." He turned to face her. "But thank you. For offering. For being here. For—" His voice broke again. "For keeping the window open. I don't know if I would have made it back without that. Without knowing you were here, waiting."

"I'll always be here." Wendy pressed her forehead to his. The gesture the Lost Boys had taught her—the ritual that meant *I see you, I'm here, we're together*.

***

Lilly Ann Morton lived in a small flat in Whitechapel.

Peter found it the next morning, following the address he had memorized from Charlie's letters. The building was modest but clean, the stairs narrow but well swept. A working woman's home, kept with pride and fierce determination.

He stood outside her door for a long time.

He had delivered death before. Had written the letters, had spoken to the families, had watched grief crack across faces like ice breaking in spring. But this was different. This was Charlie. This was the promise he had made in a trench in 1915, renewed in a dugout in 1918, carried across years of war.

Tell my mum. Tell Lilly Ann. Everything.

He knocked.

The door opened.

She looked at Peter—at his uniform, his face, the notebooks strapped to his chest—and something in her expression shifted.

She knew.

Of course she knew. The telegram had arrived weeks ago: *Regret to inform you*. But telegrams were just words. Telegrams didn't tell you how. Didn't tell you who was there. Didn't tell you the things your son said at the end.

"Hello, ma'am."

"He wrote about you. In every letter. Said you were the best friend he ever had." She stepped back from the door. "Come in. I'll make tea."

The flat was small but warm. Photographs lined the walls—Charlie as a baby, as a child, as the young man Peter had known. There was one of the Lost Boys, taken before they shipped out: all of them together, arms around each other, grinning at the camera like they had no idea what was coming.

Peter sat at the kitchen table while Lilly Ann made tea. His hands were shaking. He had faced machine guns and gas and the endless terror of No Man's Land, but this—this was harder.

"He talked about you," he said finally. "All the time. In the trenches. During the worst of it. He'd say, 'My mum would have something sharp to say about this.' Or, 'My mum would never let me give up.' You were with him. Every day. Even when you weren't."

Lilly Ann set down the teapot. Her hands were shaking too.

"Tell me," she said. "Tell me everything. How he lived. How he—" She stopped. Breathed. "How he died. I need to know."

"He survived three years." Peter's voice was raw. "Do you understand that? Three years. The Somme, Passchendaele, the German offensive. He survived all of it. He was brilliant—the smartest of all of us, the one who saw things coming, the one who kept us alive when everyone else was falling apart."

"That's my Charlie."

"He was sardonic. Always had something sharp to say. But underneath—" Peter's throat tightened. "Underneath, he was the most loyal person I ever knew. He would have died for any of us. And at the end, he almost made it. We both did. The Armistice was signed. The war was over. We just had to survive a few more hours."

"But he didn't."

"No." Peter looked at her. "A shell. November 11th. Ten fifty in the morning. Ten minutes before the ceasefire."

Lilly Ann's face crumpled. But she didn't look away. Didn't turn from the truth.

"Was he... was he in pain?"

"I don't think so. It happened fast. And I was with him—I held him. We did the thing we always did, the forehead touch, the words we said to each other." Peter pressed his own forehead, remembering.

"Still here. Still together." Lilly Ann whispered the words like a prayer. "He wrote about that. Said it was how you Lost Boys knew you were still alive. He joked it was silly, but through the lines, it meant everything to him."

"It was. It is." Peter pulled out the notebooks and set them on the table. "He wanted me to tell you everything. Not just how he died—how he lived. Who he was in the war. The things we did, the rituals we kept, the way we held on to each other in the darkness."

"Tell me."

So Peter told her.

He told her about the greenhouse where the Lost Boys had first gathered. About Billy and his photograph of Rose, the two pats every night. About The Twins and their forehead ritual, the gesture that had spread to all of them. About Tink and his wind chimes, still singing somewhere in France. About Geoffrey, who had died protecting the ledger. About Curly, who'd chosen to save the names instead of himself. About Michael, who'd wanted to write a book.

He told her about Charlie.

About his sardonic humor, his fierce intelligence, his refusal to let anyone give up. About the night he'd made Peter promise to tell Lilly Ann everything—"Not just that I died, but who I was here. What we were. The rituals and the names and the stupid forehead thing."

"He wanted you to know he was a Lost Boy," Peter said. "That he was part of something. That he was loved."

Lilly Ann was crying openly now. But she was also smiling—a broken, beautiful smile that held all the grief and all the pride of a mother who had raised a hero.

"He was my everything," she said. "My only boy. The only good thing I ever made."

"He was my brother." Peter's voice cracked. "I loved him. I still love him. And I'll carry him with me for the rest of my life."

Lilly Ann reached across the table and took Peter's hand. "Thank you. For telling me. For keeping your promise."

"I would have kept it even if I hadn't promised." Peter squeezed her hand. "That's what the Lost Boys do. We don't leave anyone behind."

***

He stayed for three hours.

They talked about Charlie—about his childhood, his dreams, the things he might have done if the war hadn't taken him. Lilly Ann showed Peter photographs he'd never seen: Charlie as a baby, chubby and serious; Charlie as a young boy, scowling at the camera, already sharp and sardonic and impossibly alive.

When Peter finally left, Lilly Ann walked him to the door.

"What will you do now?" she asked. "Now that the war is over?"

Peter thought about it. About the notebooks strapped to his chest, full of names and stories and promises. About his mother's house, full of broken soldiers learning to be whole. About the work that had started in the trenches and would continue long after the last shell fell silent.

"I'm going to tell people," he said. "About the Lost Boys. About what we did. About the ones who didn't come home and why they mattered." He touched the notebooks. "Michael wanted to write a book. He never finished it. I'm not a writer—not like he was—but I'll find a way. I'll make sure the world knows they existed."

"Charlie would like that."

"I hope so."

Lilly Ann reached up and touched his face. The gesture of a mother—not his mother, but a mother, one of the fierce women who stood in doorways trying to stop their boys from going.

"You're a good boy, Peter Panning. Charlie was lucky to have you," she said.

"I was lucky to have him."

He walked away from her door and into the London streets, carrying the weight of everyone he'd lost, carrying the light of everyone who remained.

The war was over.

The work was just beginning.

***

That night, Peter stood at the nursery window.

The prism was catching the lamplight now, scattering small rainbows across the dark glass. Outside, London was quiet—the celebrations fading, the city settling into the long work of rebuilding.

He opened the ledger to the last page. Read the entry he'd written on November 11th, in a ruined farmhouse, surrounded by soldiers who didn't know how to feel anything.

Charles "Slightly" Morton. Twenty-One. London. Sardonic, brilliant, the best friend I ever had.

He added a new line:

Told Lilly Ann. December 1918. She cried. She smiled. She said "still here, still together" like she'd known the words all along.

I think maybe she did. I think maybe all the mothers know. The ones who send their sons to war, who stand in doorways trying to stop them, who keep windows open and lights burning and refuse to give up hope.

They're Lost Boys too, in a way. Lost in the waiting. Lost in the fear. Lost in the long silence between letters.

But they find their way back. They always find their way back.

For all of us.

Peter closed the ledger.

The window was open. The prism was scattering light.

And somewhere in the darkness—in the trenches that were being filled, in the graves that were being marked, in the hearts that were learning to heal—the Lost Boys were still together.

Always.

Always.

Always.

Epilogue

Where Light Lands

London, 1928

The baby came in late November, when the leaves were falling and the world was learning to let go.

Peter stood at the nursery window—the same window, the same room, the same prism still hanging from its chain—and watched the afternoon light scatter across the walls. The rainbows moved like living things, dancing over the crib, the rocking chair, the small bundle of blankets that held his daughter.

His daughter.

The words still felt impossible. After everything—the war, the loss, the years of rebuilding—he had somehow arrived here: a father, holding a child, standing in the place where his own story had begun.

"She's perfect," his wife said. Eleanor was resting in the bed they'd brought up for the delivery, exhausted but radiant, watching Peter with eyes full of something he couldn't name. "Absolutely perfect."

"She is." Peter crossed to the crib. Looked down at the tiny face, the closed eyes, the small hands curled into fists. "She looks like you."

"She looks like herself. That's who babies look like—themselves."

"That's very philosophical."

"I'm very philosophical. It's why you married me."

Peter smiled. It was still strange, smiling—even now, ten years after the war, the expression sometimes felt borrowed, like something he was trying on rather than something that belonged to him. But Eleanor had taught him to smile again. Eleanor, and time, and the slow, patient work of healing.

He lifted the baby from the crib. Held her against his chest. Felt the impossible smallness of her, the warmth, the steady rhythm of her breathing.

"Have you thought about a name?" Eleanor asked.

Peter had been thinking about this for months. Had lists written in the margins of his notebooks—the same notebooks, still strapped to his chest sometimes when the nightmares came, still full of names that would never fade.

"Lilly Ann," he said.

Eleanor was quiet for a moment.

"After Charlie's mother?"

"After all of them." Peter looked at the baby—at Lilly Ann, his daughter, his hope. "After the fierce women who stood in doorways trying to stop their boys from going. After the ones who kept windows open and lights burning. After everyone who waited and prayed and refused to give up."

"It's beautiful."

"It's true." Peter touched his daughter's face. "They saved us, you know. The mothers. The Lilly Anns. We couldn't have survived without them."

"You survived because of each other. You and the Lost Boys."

"We survived because they gave us something to come home to." Peter's voice was soft. "A window. A light. A name that meant we weren't forgotten. That's what Lilly Ann did for Charlie. That's what my mother did for me. And now..." He looked at the prism, scattering rainbows across his daughter's face. "Now I'm going to do it for her."

***

Wendy appeared in the doorway an hour later.

She was older now—truly old, her hair white, her face lined with the map of a life well-lived, and the added grief of being a widow. But her eyes were the same. Still watching. Still waiting. Still ready to catch whatever needed catching.

"May I?" she asked.

Peter placed Lilly Ann in his mother's arms.

Wendy held the baby the way she'd held a thousand broken soldiers, a thousand lost boys, a thousand children who needed someone to believe they could find their way home. With tenderness. With patience. With the particular strength of women who had learned that love was the only light that never went out.

"Lilly Ann," Wendy said. "That's her name?"

"Yes."

"For Charlie's mother."

"For all of them."

Wendy nodded slowly. Something moved across her face—grief and joy and pride, all tangled together, impossible to separate.

"She would have liked that," Wendy said. "Lilly Ann Morton. She came to visit, you know, a few years ago. After you'd told her about Charlie. She wanted to see the window."

"I didn't know that."

"She stood right where you're standing. Watched the prism scatter light. And she said—" Wendy's voice cracked. "She said, 'So this is where they came from. The boys who saved my Charlie. This is where the light started.'"

Peter felt tears sliding down his cheeks. He didn't wipe them away.

"What happened to her? Lilly Ann Morton?"

"She died. Two years ago. Peacefully, in her sleep." Wendy looked down at the baby. "But she knew about this. Before she went. I wrote to tell her you were married, that Eleanor was expecting. She wrote back and said..." Wendy smiled through her tears. "She said to name the baby Lilly Ann. If it was a girl. She said Charlie would have liked that."

"She knew?"

"She hoped. Mothers always hope." Wendy pressed her forehead to the baby's. The gesture the Lost Boys had taught her. The gesture that had spread from The Twins to all of them, from the trenches to the nursery, from death to life.

***

Peter wrote to Rose Whitmore that evening. She was married now—had married a man from her village, a quiet schoolteacher who had lost his arm at Verdun, who understood loss in the way that only survivors could. They had children: two boys, a girl.

Dear Rose,

I wanted you to know we had a daughter today. We named her Lilly Ann after Charlie's mother, and after all the women who waited and hoped and kept windows open.

I think of Billy often. Of the wedding you never had, the house you never built, the six children he wanted to fill it with. I think of him saying "after" like it was a prayer, believing so hard in a future he never got to see.

But here's what I've learned: "after" came anyway. Not the way he imagined. Not with him in it. But it came. There are children in the world now—yours, mine—who exist because we survived, because we kept going, because we refused to let the darkness win.

That's Billy's legacy. That's what he gave us. Not a wedding, not a house, not six children of his own—but hope. The stubborn, relentless belief that there would be an "after." And there is. There is.

Lilly Ann is sleeping as I write this. The prism in the window is scattering rainbows across her crib. And I keep thinking: this is what light does. It doesn't disappear. It just finds new places to land.

The boys left there in Neverland will never grow old.

Peter

***

The years passed.

Lilly Ann grew. First steps, first words, first day of school. Peter watched her with the intensity of a father who had seen too much death to take life for granted—every smile a miracle, every laugh a gift, every ordinary moment precious beyond measure.

He told her stories. That was what fathers did, after all—they told stories. But Peter's stories were different. They were true.

He told her about the Lost Boys.

Not all at once—she was too young for the full weight of it, too innocent to carry the knowledge of trenches and gas and boys dying in the mud. But in pieces, in fragments, in the small truths that children could hold.

He told her about Billy, who had believed in "after" so hard you could almost see it. About The Twins, who had loved each other so fiercely they couldn't exist apart. About Geoffrey, who had been young and brave and had died protecting something bigger than himself.

He told her about Curly, the quiet guardian. About Michael, who had wanted to write a book. About Tink, who'd made wind chimes from shell casings, who turned destruction into music.

And he told her about Charlie.

"Uncle Slightly," Lilly Ann called him, though they had never met. "Tell me about Uncle Slightly, Daddy."

So Peter told her. About the sardonic humor, the fierce intelligence, the boy who had pretended not to care but cared more than anyone. About Lilly Ann Morton, the mother who had given Charlie everything—her strength, her stubbornness, her refusal to give up.

"That's who you're named for," Peter said. "Not just her. All of them. The mothers and the sisters and the girls who waited. The ones who kept the light burning."

"Like Nana Wendy?"

"Exactly like Nana Wendy."

Lilly Ann considered this with the gravity of a seven-year-old contemplating deep truths.

"So I'm named after the light," she said.

"Yes." Peter felt something break and mend in his chest. "You're named after the light."

***

The wind chimes hung in the garden.

Peter had made them himself—not from shell casings, not anymore, but from bits of brass and copper, from fragments of things that had survived the war. They sang in the breeze, high and bright, the same music Tink had coaxed from destruction.

Lilly Ann loved them. She would sit in the garden for hours, listening, watching them sway.

"They're talking," she announced one day. "The wind chimes. They're saying something."

"What are they saying?"

"'Still here.'" She looked at her father with eyes that saw more than a child should. "That's what they're saying, Daddy. 'Still here. Still together.'"

Peter knelt beside her and took her small hand in his.

"That's exactly what they're saying."

"Who taught you that? The words?"

"The Lost Boys. A long time ago, before you were born. We used to say it to each other every day. Every night. Whenever things got dark."

"What does it mean?"

Peter thought about the question. About the trenches, the dugouts, the moments when the words were all they had. About Billy saying them before the shell came. About The Twins pressing their foreheads together in the chaos. About Charlie, whispering them at the end.

"It means we're connected," he said finally. "No matter what happens. No matter how far apart we are, or how much time passes, or whether we're..." He stopped. "Whether we're here or not. The people we love, the people who love us—we're always together. In the words. In the memories. In the light that never goes out."

Lilly Ann squeezed his hand.

The wind chimes sang.

The rainbows scattered.

And in a garden in London, a father and daughter held on to each other and listened to the music the Lost Boys had left behind.

***

Peter Panning died in 1962.

He was sixty-five years old—older than any of the Lost Boys had lived to be, older than he had ever expected to become. He had spent the years since the war telling stories, keeping windows open, carrying the names of the ones who hadn't made it back.

He had written a book—not a proper book, not the kind Michael had dreamed of, but a collection of entries from the ledger. *The Lost Boys of the Great War*, it was called. It was published in 1938, just before another war came to swallow another generation.

People read it. That was what mattered. People read it, and they remembered.

Lilly Ann was with him at the end.

She was in her forties now, a mother herself, with children who called Peter "Papa" and listened to his stories with the wide-eyed wonder of the young. She sat beside his bed in the house on Wimpole Street—the same house, the same nursery, the same window still open after all these years—and held his hand.

"Tell me again about Uncle Slightly," she said. It was what she always said when she wanted to comfort him. The story he loved most, even though it hurt the most too.

"Charlie Morton," Peter whispered. His voice was thin, fading, but the words were still there. "Slightly. The best friend I ever had."

"What would he say? If he could see you now?"

Peter smiled. A real smile—not borrowed, not practiced, but earned through sixty years of learning how to live.

"He'd say I got old," Peter said. "He'd say I got soft. He'd make some sardonic comment about how I was always too sentimental for my own good."

"And then?"

"And then he'd press his forehead to mine. And he'd say the words." Peter closed his eyes. "Still here."

"Still together," Lilly Ann whispered.

"Always."

Peter Panning died with the light from the prism scattered across his face, with his daughter's hand in his, with the wind chimes singing in the garden below.

He was buried next to his mother and father. The grave was simple—just a stone, just a name, just a few words carved into the marble:

Peter Panning

1897–1962

A Lost Boy

* * *

The window is still there.

In the house on Wimpole Street, in the nursery where it all began, the prism still hangs from its chain. Lilly Ann checks on it every year, on November 11th, at eleven o'clock—the exact moment the guns stopped, the exact moment her namesake's son died in her father's arms.

She opens the window. Watches the light scatter.

And she says the words.

The Lost Boys answer.

They answer in the wind chimes singing in the garden. In the rainbows moving across the walls. In the photographs on the mantelpiece, faded now but still smiling, still young, still impossibly alive.

Billy is there, gap-toothed, holding Rose's photograph. The Twins are there, hands clasped. Geoffrey is there, too young, too eager, burning to belong. Curly is there, solid and silent, watching the door. Michael is there, notebook in hand, recording everything. Tink is there, humming tunelessly, making music from nothing.

And Charlie is there. Slightly. Standing at Peter's shoulder, sardonic and sharp, the best friend anyone could ask for.

The light doesn't disappear. It just finds new places to land. In the children who carry the names forward. In the stories that pass from generation to generation. In the windows that stay open, waiting for the lost ones to find their way home.

That's what the Lost Boys taught us.

That's what they died to prove.

That even in the darkest places—in the trenches and the mud and the endless ticking of the crocodile—the light survives. The names survive. The love survives.

Still here.

Still together.

Always.

A Note on History and Imagination

Wendy's Home for Lost Boys is a work of fiction that draws deeply on historical reality. Readers may wonder where the line falls between documented fact and narrative invention.

WHAT IS HISTORICALLY ACCURATE

The scale of loss. Nearly 900,000 British and Commonwealth soldiers died in the First World War. Millions more were wounded, many catastrophically. The youngest confirmed British soldier killed was just fourteen years old.

Underage enlistment. Approximately 250,000 boys under the legal enlistment age of nineteen served in the British Army during WWI. Many were as young as fourteen or fifteen. Recruitment officers, under pressure to meet quotas, often looked the other way when boys lied about their ages. Some of these children are buried in Commonwealth War Graves cemeteries across France and Belgium, their headstones recording ages that were themselves lies.

The Christmas truce of 1914. The informal ceasefire described in this novel occurred along portions of the Western Front on Christmas Eve and Christmas Day, 1914. British and German soldiers left their trenches, exchanged gifts, sang carols, and in some cases played football in No Man's Land. The truce was not officially sanctioned and was actively suppressed by military authorities in subsequent years.

The Battle of the Somme. The first day of the Somme offensive, July 1, 1916, remains the bloodiest single day in British military history. Nearly 60,000 British soldiers became casualties in a single morning, roughly 19,000 of them killed. The battle continued for four and a half months, resulting in over one million total casualties on all sides for a gain of approximately six miles.

Deaths on the final day. Fighting continued until 11:00 a.m. on November 11, 1918, despite the Armistice being signed hours earlier. An estimated 11,000 men were killed, wounded, or went missing on the war's final morning. Some commanders ordered attacks in the final minutes, believing a strong finish would look better in regimental histories. The last British soldier killed in the war, George Edwin Ellison, died at 9:30 a.m.—ninety minutes before the ceasefire.

Native American service. Approximately 12,000 Native Americans served in the U.S. military during World War I, despite the fact that roughly one third were not recognized as U.S. citizens. They volunteered at extraordinarily high rates and suffered disproportionate casualties—about 5 percent of Native American soldiers were killed, compared to 1 percent of U.S. forces overall. The Choctaw Code Talkers of the 142nd Infantry, 36th Division, pioneered the use of Native languages for military communications—a practice the Germans never broke and one that laid the groundwork for the famous Navajo code talkers of World War II. Native American soldiers fought for a nation that did not grant them citizenship until the Indian Citizenship Act of 1924, passed in part because of their wartime service.

What Is Imagined

The Panning family and all named characters in this novel are fictional.

The connection to J.M. Barrie's *Peter Pan* is, of course, an imaginative reimagining. The real *Peter Pan* premiered on December 27, 1904, at the Duke of York's Theatre in London. Barrie's encounters with the Panning family are invented, though his lifelong preoccupation with lost children and his complicated relationships with the Llewelyn Davies boys are well documented.

The "ledger" that Peter keeps—the record of names and stories—is inspired by the real work of graves registration units and the Imperial (now Commonwealth) War Graves Commission, whose mission to identify, bury, and commemorate every fallen soldier regardless of rank remains one of the most remarkable humanitarian undertakings in history.

The characters of Thomson and Thompson—"The Twins"—are fictional, but they are inspired by the real Choctaw soldiers who served in the 36th Division. Their names honor documented Code Talkers including Solomon Louis, Mitchell Bobb, Ben Carterby, Tobias Frazier, and Joseph Oklahombi, among others. The decision to make them cousins from Oklahoma rather than biological twins reflects the historical reality of Choctaw enlistment from Indigenous Territory.

A Note From the Author

This novel is my attempt to honor the lost boys of the First World War—not to glorify war, but to remember the children who fought it. It is also my attempt to honor the caretakers whose work was rarely recognized but absolutely essential.

The Peter Pan mythology woven through this story is inspired by J.M. Barrie's beloved work. In 1929, Barrie gave the copyright to *Peter Pan* to Great Ormond Street Hospital for Children, ensuring that his story about the boy who wouldn't grow up would forever support children fighting to grow up at all. That gift continues today.

If this novel has moved you, the resources below are a place to start.

—Sara Madden

Resources for Readers, Teachers, and Librarians

Understanding the First World War

For Young Readers (Ages 10–14)

War Horse by Michael Morpurgo – A beloved novel about a horse's journey through WWI, offering an accessible entry point to understanding the war's impact on both humans and animals.

Private Peaceful by Michael Morpurgo – The story of two brothers in the trenches, dealing with themes of underage enlistment, shell shock, and military justice. An essential YA companion to the themes in this novel.

Soldier Dog by Sam Angus – A story about a young soldier and the dog he trains for war service, exploring themes of loyalty, loss, and healing.

The War That Saved My Life by Kimberly Brubaker Bradley – Set during WWII but excellent for understanding child evacuees, trauma, and recovery—themes that echo across both world wars.

For Older Readers (Ages 14+)

All Quiet on the Western Front by Erich Maria Remarque – The classic novel of WWI from a German soldier's perspective—essential for understanding that "lost boys" existed on all sides.

Birdsong by Sebastian Faulks – A literary exploration of the war's impact on soldiers and the women who loved them.

Regeneration trilogy by Pat Barker – A masterful fictional exploration of shell shock and the psychiatric treatment of WWI soldiers, based on real historical figures.

Testament of Youth by Vera Brittain – A memoir by a VAD nurse who lost her fiancé, brother, and friends to the war—essential reading for understanding the caretakers' experience.

BOY SOLDERIES OF WWI

Books

Boy Soldiers of the Great War by Richard van Emden – A comprehensive history of underage soldiers, including firsthand accounts and photographs. Essential reading for understanding how children ended up in the trenches.

Teenage Tommy: Memoirs of a Cavalryman in the First World War by Richard van Emden – The story of Ben Clouting, who enlisted at fifteen and survived the entire war.

Too Young to Die: Boy Soldiers of the First World War by Dan Snow and Taff Gillingham – An accessible overview with photographs and personal stories.

Websites and Digital Resources

Imperial War Museum: Lives of the First World War www.livesofthefirstworldwar.iwm.org.uk – A digital memorial with records of those who served. Teachers can use this for research projects.

The Long, Long Trail www.longlongtrail.co.uk – Comprehensive information about the British Army in WWI, including unit histories and soldier experiences.

BBC History: World War One www.bbc.co.uk/history/worldwars/wwone – Articles, videos, and interactive resources suitable for classroom use.

Shell Shock and PTSD: Understanding the History

Broken Men: Shell Shock, Treatment and Recovery in Britain 1914-1930 by Fiona Reid – An academic but accessible history of how shell shock was understood and treated.

Shell Shock: Traumatic Neurosis and the British Soldiers of the First World War by Peter Leese – Explores the medical, military, and social responses to psychological trauma.

Shattered Nerves: Doctors, Patients, and Depression in Victorian England by Janet Oppenheim – Provides context for how mental illness was understood before WWI.

For Young Readers

Finding Someplace by Denise Lewis Patrick – While set during Hurricane Katrina, this novel deals with trauma and recovery in ways young readers can understand.

The Memory of Light by Francisco X. Stork – A YA novel about depression and recovery that can help young readers understand the psychological wounds discussed in historical context.

Modern Resources

Combat Stress (UK) www.combatstress.org.uk – The UK's leading veterans' mental health charity, continuing the work of supporting "lost boys" today.

Veterans Crisis Line (US) www.veteranscrisisline.net – 1-800-273-8255, Press 1 For any veteran or service member in crisis.

THE CHRISMAS TRUCE OF 1914

Books

Silent Night: The Story of the World War I Christmas Truce by Stanley Weintraub – The definitive account of the spontaneous truces that broke out along the Western Front.

Meeting in No Man's Land: Christmas 1914 and Fraternisation in the Great War by Marc Ferro – A scholarly examination of the truce and its aftermath.

For Young Readers

War Game by Michael Foreman – A picture book telling the story of the Christmas truce, suitable for younger readers.

Truce: The Day the Soldiers Stopped Fighting by Jim Murphy – An accessible account for middle-grade readers.

Films

Joyeux Noël (2005) – A dramatization of the Christmas truce, suitable for older teens. (Rated PG-13)

THE CARETAKERS: NURSES, VADS, AND RED CROSS WORKERS

Books

Nurses of Passchendaele: Caring for the Wounded of the Ypres Campaigns 1914-1918 by Christine E. Hallett – The experiences of nurses on the Western Front.

Veiled Warriors: Allied Nurses of the First World War by Christine E. Hallett – A comprehensive history of military nursing in WWI.

Not So Quiet: Stepdaughters of War by Helen Zenna Smith – A novel based on the diaries of an ambulance driver, offering a woman's perspective on the war.

Testament of Youth by Vera Brittain – Essential reading—a firsthand account of nursing during WWI and its lasting impact.

For Young Readers

The War Nurse by Tracey Enerson Wood (YA edition available) Historical fiction about nurses in WWI.

Organizations to Learn More

British Red Cross Museum and Archives www.redcross.org.uk/about-us/our-history – Archives and educational resources about Red Cross work during WWI.

The Florence Nightingale Museum www.florence-nightingale.co.uk – While focused on an earlier era, provides context for the nursing profession.

NATIVE AMERICANS IN WWI

Understanding the History

Native Americans have served in the U.S. military at higher rates than any other ethnic group, a tradition that began long before they were recognized as citizens. During WWI, they fought for a country that had taken their land, suppressed their languages, and denied them the vote. Their service is a story of profound complexity: warrior traditions, love of homeland, and hope that sacrifice might finally earn recognition.

Books

The First Code Talkers: Native American Communicators in World War I by William C. Meadows – The definitive account of WWI code talkers, covering all known Native American code talkers including the Choctaw, Cherokee, Comanche, Osage, and Sioux nations. Essential reading for understanding this overlooked chapter of military history.

Why We Serve: Native Americans in the United States Armed Forces (Smithsonian Books, 2020) – A 240-page companion to the National Native American Veterans Memorial, exploring 250 years of Native American military service.

American Indians in World War I: At War and at Home by Thomas A. Britten – Examines the broader context of Native American participation, including the home front and the fight for citizenship.

The Choctaw Code Talkers

In October 1918, Colonel Alfred Bloor of the 142nd Infantry overheard two Choctaw soldiers conversing and realized that their language—never written down and unknown to outsiders—could serve as an unbreakable code. Eight Choctaw soldiers were initially deployed, transmitting messages that allowed American forces to capture German positions with complete surprise. The practice expanded to include soldiers from the Cherokee, Comanche, Osage, and other nations.

The Choctaw language lacked military terminology, so the soldiers invented code within the code: "stone" for grenade, "big gun" for artillery, "fast shooting gun" for machine gun. The Germans never broke it.

The Citizenship Paradox

Native American soldiers fought for a country that did not recognize them as citizens. Many were drafted despite their non-citizen status; others volunteered, hoping their service would earn recognition. After the war, a 1919 act offered citizenship to Native veterans—but the process was so complicated that most simply went home. Full citizenship for all Native Americans did not come until 1924, and voting rights in some states were denied until the 1940s.

WILD WEST SHOWS AND NATIVE AMERICANS IN EUROPE

Understanding the History

Between 1883 and the outbreak of World War I, Wild West shows brought hundreds of Native American performers to Europe. Buffalo Bill's Wild West toured England eight times, performing for Queen Victoria at her Golden Jubilee in 1887 and drawing over 2.5 million spectators in London alone. The Miller Brothers 101 Ranch Wild West, based in Oklahoma, toured Europe from 1908 to 1914, performing for King George V and Queen Mary of England.

These shows employed primarily Lakota Sioux performers, but Native Americans from many nations—including those from

Indian Territory (present-day Oklahoma)—traveled with various Wild West productions. Performers signed seasonal contracts and were paid wages, though conditions varied widely between shows. Some performers stayed behind when shows disbanded or moved on, stranded in foreign cities. Others chose not to return. The 1895 Montana's Wildest Wild Show, for instance, disbanded in Cincinnati, leaving its Cree performers stranded. When the Miller Brothers' show was caught in Europe at the outbreak of World War I, the company scrambled to evacuate performers through Norway, though some Oglala Sioux performers arrested in Germany as suspected spies were never seen again.

For Native Americans, the shows represented a complex bargain. They could earn wages, travel the world, and continue practicing cultural traditions—dances, regalia, horsemanship—that were being actively suppressed on reservations. Historian Vine Deloria Jr. argues that Show Indians spent their time "playing Indian" as a form of cultural resistance: "Perhaps they realized in the deepest sense that even a caricature of their youth was preferable to a complete surrender to the homogenization that was overtaking American society."

The Choctaw Nation, based in southeastern Oklahoma, had strong ties to the broader Wild West show circuit through geographic proximity to the Miller Brothers' 101 Ranch and other Oklahoma-based productions. While most documented Show Indians were Plains tribes, the shows drew performers from across Indian Territory's diverse nations.

Books

Wild West Shows and the Images of American Indians, 1883-1933 by L.G. Moses – The definitive scholarly account of Native American performers in Wild West shows. Moses examines Show Indians from their own perspective, arguing that their participation helped preserve cultural practices during decades of forced assimilation. Essential reading for understanding this complex chapter of Native American history.

The Real Wild West: The 101 Ranch and the Creation of the American West by Michael Wallis – Award-winning history of the Miller Brothers' 101 Ranch and its influential Wild West show. Wallis documents the show's European tours, its cast of performers (including Native Americans, African American cowboys, and cowgirls), and its role in creating the mythology of the American West. Winner of the Spur Award, Western Heritage Award, and Oklahoma Book Award.

Indians in Unexpected Places by Philip J. Deloria – Broader cultural history examining Native Americans in spaces where white audiences did not expect them, including European tours. Excellent for understanding how Native performers navigated expectations and stereotypes while asserting their own identities.

The Fabulous Empire: Colonel Zack Miller's Story by Fred Gipson – First-person account from one of the Miller Brothers, documenting the 101 Ranch Wild West's adventures, including its European tours and the chaos of evacuating performers at the outbreak of World War I.

The 101 Ranch by Ellsworth Collings and Alma Miller England – The foundational history of the ranch and its show, written with family cooperation. Includes details on Native American performers and their relationships with the Miller family.

Websites and Digital Resources

Buffalo Bill Center of the West www.centerofthewest.org – Extensive archives on Wild West shows, including photographs of Native American performers. The McCracken Research Library houses primary source materials including route books, photographs, and correspondence.

Wičhóoyake kiŋ aglí Project (Oglala Lakota College/Clemson University) – A collaborative research project documenting Lakota ancestors who performed in Buffalo Bill's Wild West Show, gathering stories, photographs, and artifacts from descendants. Funded by the National Archives.

The William F. Cody Archive codyarchive.org – Digital archive documenting the life of Buffalo Bill, including extensive materials on his Wild West shows and Native American performers. Includes photographs, programs, and route books.

Oklahoma Historical Society: Wild West Shows and Performers www.okhistory.org/publications/enc/entry.php?entry=WI010 – Encyclopedia entry on Oklahoma's role in Wild West show history, including the Miller Brothers 101 Ranch and its connections to Native American nations in Indian Territory.

Showmen's Museum www.showmensmuseum.org – Preserves the history of traveling shows including Wild West productions. Includes materials on the 101 Ranch Wild West Show.

The Irony of the Show Indian

Thomson and Thompson arrive in London as performers in a world that wanted to watch them play "Indian" for entertainment—then asked them to die in its war. This was not unique. Native American soldiers in World War I fought for a country that did not recognize them as citizens, having first been displayed for European audiences as exotic spectacles. The journey from Show Indian to soldier represents one of the strangest and most poignant transformations of the war.

The Wild West shows taught European audiences a fantasy of the American frontier. The war taught everyone that the fantasy was over.

Websites and Digital Resources

Choctaw Nation of Oklahoma: Code Talkers www.choctawnation.com/about/history/code-talkers – History and profiles of the original Code Talkers.

National WWI Museum: America's First Code Talkers www.theworldwar.org/learn/about-wwi/americas-first-code-talkers – Educational resources including a virtual reality experience recreating Code Talker communications.

National Archives: Code Talkers www.archives.gov/research/native-americans/military/code-talkers – Primary source documents and historical records.

Smithsonian National Museum of the American Indian www.americanindian.si.edu – "Why We Serve" exhibition and extensive archives on Native American military history.

Honoring the Dead: Burial Practices and The Graves Registration Service

Books

The Unending Vigil: A History of the Commonwealth War Graves Commission by Philip Longworth – The definitive history of how the dead of WWI were buried and commemorated.

Empires of the Dead: How One Man's Vision Led to the Creation of WWI's War Graves by David Crane – The story of Fabian Ware and the creation of the Imperial War Graves Commission.

Known Unto God: The Unidentified Dead of the Great War by Neil Hanson – A moving exploration of the soldiers buried as "Known Unto God"—unnamed but not forgotten.

Visiting the Memorials

Commonwealth War Graves Commission www.cwgc.org – Search for graves and memorials, plan visits, and learn about the commission's ongoing work. An invaluable resource for students researching family members who served.

Thiepval Memorial, France – The memorial to the missing of the Somme, bearing the names of over 72,000 soldiers with no known grave.

Menin Gate, Ypres, Belgium – The memorial to the missing of the Ypres Salient. Every evening at 8 p.m., the Last Post is sounded—a tradition that has continued since 1928 (except during German occupation in WWII).

J.M. Barrie and the Gift of Peter Pan

In 1929, J.M. Barrie gave the copyright to Peter Pan to Great Ormond Street Hospital for Children in London. This extraordinary gift has generated millions of pounds for the hospital, supporting children's healthcare for nearly a century.

About the Gift

Great Ormond Street Hospital: The Peter Pan Connection www.gosh.org/about-us/peter-pan – Learn about Barrie's gift and its continuing impact.

Books About Barrie and Peter Pan

J.M. Barrie and the Lost Boys by Andrew Birkin – The definitive biography of Barrie and his relationship with the Llewelyn Davies boys who inspired Peter Pan.

Neverland: J.M. Barrie, the Du Mauriers, and the Dark Side of Peter Pan by Piers Dudgeon – A more complex exploration of Barrie's life and inspirations.

Peter Pan by J.M. Barrie – Read the original—the play (1904) and the novel *Peter and Wendy* (1911). Notice how Barrie's understanding of lost boys, of the space between childhood and adulthood, resonates with the themes of this novel.

Supporting the Legacy

Great Ormond Street Hospital Children's Charity www.gosh.org – Consider making a donation in honor of the lost boys—both the ones in Barrie's story and the ones in ours.

VETERANS SUPPORT: THEN AND NOW ETERANS SUPPORT: THEN AND NOW

Historical Resources

Broken Men: Shell Shock, Treatment and Recovery in Britain 1914-1930 by Fiona Reid – Understanding how veterans were (and often weren't) supported after WWI.

The Pity of War by Niall Ferguson – Includes analysis of how societies dealt with returning veterans.

The Ex-Servicemen's Welfare Society

Wendy's Home for Lost Boys is inspired by the real homes and hostels established for veterans after WWI. The Ex-Servicemen's Welfare Society (now part of the Royal British Legion) operated hostels for homeless and struggling veterans.

The Royal British Legion www.britishlegion.org.uk – Continuing to support veterans today.

Modern Veterans' Organizations

Help for Heroes (UK) www.helpforheroes.org.uk

Wounded Warrior Project (US) www.woundedwarriorproject.org

The Veterans Project www.veteransproject.org – Oral history project preserving veterans' stories.

For Teachers and Librarians

Curriculum Connections

This novel connects to:

- World War I history
- British literature and the modernist period
- Psychology and mental health awareness
- Social studies: citizenship, service, and sacrifice
- Media literacy: examining how war is portrayed vs. experienced

Discussion Questions

1. How does the novel transform the Peter Pan mythology? What does Neverland represent in this context?

2. Peter writes names to fight against erasure. Why is naming—and being named—so important to the Lost Boys?

3. How do Wendy's and Peter's experiences of the war parallel and differ from each other?

4. The novel portrays "Hook" not as a person but as something else. What is Hook in this story? How does this change your understanding of villains and heroism?

5. Why do you think soldiers continued to die on the last day of the war, after the Armistice was signed? What does this tell us about the nature of war and those who direct it?

6. How do the "caretakers" in the novel—nurses, mothers, gravediggers—contribute to the survival of the lost boys? Why might their stories be less well-known than soldiers'

stories?

7. The prism appears throughout the novel. What does it symbolize, and how does its meaning develop?

8. Peter struggles with his "shadow" and with "Hook" inside his mind. How does the novel portray psychological trauma? How might this help modern readers understand PTSD?

9. Thomson and Thompson are Choctaw cousins who fight for a country that doesn't recognize them as citizens. Why do you think Native Americans volunteered at such high rates despite this? How does their story complicate our understanding of patriotism and belonging?

Classroom Activities

Names Project – Have students research a real soldier from WWI using the Commonwealth War Graves Commission database. Write a "ledger entry" for that soldier, imagining their story based on available facts.

Letters Home – Write a pair of letters—one from a soldier at the front, one from a family member at home—exploring what can and cannot be said across the distance of war.

Caretaker Research – Research one of the "caretaker" roles from WWI: VAD nurse, Red Cross worker, gravedigger, chaplain, etc. Present findings to the class.

Memorial Design – Design a memorial for the lost boys of WWI. Consider: What should be remembered? How can a memorial honor both the dead and the survivors?

Code Talker Research – Research the Choctaw Code Talkers or another group of Native American soldiers from WWI. How did their languages become military assets after being suppressed in boarding schools? What does this reveal about the complexity of Native American military service?

Museums and Memorials to Visit

United Kingdom

Imperial War Museum, London www.iwm.org.uk – Extensive WWI galleries, including immersive trench experience.

Imperial War Museum North – Manchester sister museum with rotating exhibitions.

National Army Museum, London www.nam.ac.uk – Collections spanning British Army history.

France and Belgium

In Flanders Fields Museum, Ypres, Belgium www.inflandersfields.be – Interactive museum in the medieval Cloth Hall.

Thiepval Visitor Centre, France www.cwgc.org/visit-us/find-cemeteries-memorials/cemetery-deta

ils/150700/thiepval-anglo-french-cemetery/ – Memorial to the missing of the Somme.

Tyne Cot Cemetery, Belgium – The largest Commonwealth War Graves Commission cemetery in the world.

UNITED STATES

National World War I Museum and Memorial, Kansas City www.theworldwar.org – America's official WWI museum.

National Native American Veterans Memorial, Washington, D.C. www.americanindian.si.edu/visit/washington/nnavm – Located on the grounds of the National Museum of the American Indian. The first national landmark to honor Native American, Alaska Native, and Native Hawaiian veterans. Designed by Harvey Pratt (Cheyenne and Arapaho), the memorial features an elevated steel circle on a carved stone drum, with water, fire, and lances for prayer cloths. Open 24 hours, free admission.

Tvshkahomma Capitol Museum, Tuskahoma, Oklahoma www.choctawnation.com/about/culture/tvshkahomma-capitol-museum – The Choctaw Nation's museum featuring a memorial exhibit honoring the Choctaw Code Talkers with medals, personal effects, and descendant interviews. Free admission.

Choctaw Cultural Center, Calera, Oklahoma www.choctawculturalcenter.com – Over 100,000 square feet of exhibitions exploring Choctaw history and culture.

A Final Word

The lost boys of the First World War have been gone for a century now. The last veteran died in 2012. Soon, there will be no one left who remembers them firsthand.

But we can still remember them in other ways.

We can read their letters. Visit their graves. Speak their names.

We can learn from their stories. Not just the glory, but the cost. Not just the victory, but the price.

And we can keep the window open.

For all the lost boys, wherever they are.

For all the lost boys, still waiting to be found.

Also by Sara Madden

Hidden Histories of War Series

Known Unto Us: The History of War's Caretakers
The Healer's Burden: Medica, Nurses, and Battlefield Medicine

Hidden Histories of Women Series

The Blame: A History of Heresy, Genius, and the Violence of Erasure
Women and Culture: The Americas and Beyond – Hidden Narratives of Power, Preservation, and Reclamation

Young Adult Cross-Over Rewoven Fairy Tale Books

The Weaver and the Wolf: A Rewoven Fairy Tale of Little Red Riding Hood (companion to *The Blame: A History of Heresy, Genius, and the Violence of Erasure*)

Wendy's Home for Lost Boys: A Rewoven Fairy Tale of Peter Pan (companion to *Known Unto Us: The History of War's Caretakers)*

YOUNG ADULT BOOKS

Who We Are: Heroes & Bullies

EARLY CHAPTER READER BOOKS

The Adventures of Tallulah Froom Un-Book One
The Adventures of Tallulah Froom Un-Book Two

CHILDREN'S PICTURE BOOKS

Jahan Akua
Jun Anzu
Lucy's Umbrella
Pauly & Fluff's Day at the Beach
The Elf Who Didn't Believe in Children
The Kind Kindergarten Class
The Prince's Dress Dilemma
Too

About the Author

Sara Madden is a writer, historian, and anthropologist whose work centers on forgotten lives, erased narratives, and the people history asks to bear impossible burdens in silence.

She is the author of multiple narrative nonfiction and myth-reimagining works across her *Hidden Histories* and *Rewoven Fairy Tales* series, which explore themes of memory, care, trauma, and moral responsibility through both historical and symbolic lenses. Her writing blends rigorous research with lyrical storytelling, giving voice to those often reduced to footnotes—or left out of the record entirely.

Drawing on her academic background in history and anthropology, as well as her professional experience working with humanitarian organizations and vulnerable communities, Madden approaches storytelling as an act of witnessing. Her work consistently asks not only *what happened*, but *who remembered—and who kept watch*.

She lives in Utah, where she continues to write about war, childhood, care, and the enduring human instinct to leave the window open.

www.ingramcontent.com/pod-product-compliance
Lightning Source LLC
LaVergne TN
LVHW100728061126
851857LV00011B/440